Space
Oddities

Space
Oddities

**The Mysterious Anomalies Challenging
Our Understanding of the Universe**

Harry Cliff

Doubleday ⚓ New York

Jacket images: (Earth) fotograzia; (Moon) Kativ; (infinity
symbol) Liyao Xie; (star cluster) Javier Zayas Photography;
(starry sky) Justin Helmick / EyeEm; all Getty Images
Jacket design by Michael J. Windsor

Library of Congress Cataloging-in-Publication Data
Names: Cliff, Harry (Harry Victor), author.
Title: Space oddities : the mysterious anomalies challenging
our understanding of the universe / Harry Cliff.
Description: New York : Doubleday, [2024] |
Includes bibliographical references.
Identifiers: LCCN 2023029383 (print) | LCCN 2023029384 (ebook) |
ISBN 9780385549035 (hardcover) | ISBN 9780385549042 (ebook)
Subjects: LCSH: Cosmology—Popular works. |
Particles (Nuclear physics)—Popular works.
Classification: LCC QB982 .C55 2024 (print) |
LCC QB982 (ebook) | DDC 523.1—dc23/eng/20231016
LC record available at https://lccn.loc.gov/2023029383
LC ebook record available at https://lccn.loc.gov/2023029384

MANUFACTURED IN THE UNITED STATES OF AMERICA
1st Printing

To my colleagues on the LHCb experiment,
whose tireless search for answers inspired this story

The most exciting phrase to hear in science, the one that heralds new discoveries, is not "Eureka!" (I found it!) but "That's funny . . ."

—Isaac Asimov

Contents

Space
Oddities

Prologue

December 2014, Antarctica

Twenty-three miles above Antarctica, a huge, translucent orb hangs high in the thin, freezing air. Up here, in the tenuous stratosphere, the balloon has swollen to monstrous proportions, a bloated sphere the size of a football stadium, floating free on the polar winds. Below, suspended on a long cable, hangs a peculiar object, tiny by comparison, reminiscent of some oversized speaker system, its forty-eight radio horns gleaming white in the perpetual daylight.

Far, far below stretches the vast expanse of the Antarctic ice sheet, bright, cold, and remote. From its lofty vantage point, the ANITA antenna watches over one million cubic kilometers of ice, waiting patiently for a signal. It is a week since the balloon took flight from the edge of the ice sheet and began its slow orbit of the continent on the circumpolar winds, and since then it has seen nothing out of the ordinary.

Until now.

Imperceptible at this distance, a blue light flickers deep in the ice. Microseconds later, a fleeting blast of radio noise flashes upward over the antenna and vanishes into space. It's all over in an instant, but it will be four years until the team of scientists who launched this strange device into the Antarctic skies realizes the significance of what has just occurred.

January 25, 2021, London, U.K.

"Okay, guys, this is it." Mitesh Patel shifts nervously in his office chair. In front of him, the faces of six of his closest colleagues stare back at him from his computer screen. No one speaks, but the tension is palpable. He notices Kostas resting his head between two closed fists, knuckles pressed hard into either temple, his lips tightened with anticipation. Or is it anxiety?

For a moment, the only sound is the gentle pattering of winter drizzle on the darkened window to his left. He realizes that he is shaking. Paula's voice breaks the silence. "Okay, the fit is finished." Her eyes dart from left to right as she scans the printout on her terminal, then, perhaps, the faint flicker of a smile. "Are you ready to hear it?"

His heart pounds in his chest. Particle physics is rarely a high-stakes business, but tonight is one of those rarest of moments on which the future of the subject can turn. The anomaly in their analysis has been building for the best part of a decade, as the LHCb experiment has recorded trillions upon trillions of particle collisions in a cavern deep beneath the French countryside. Tonight is the culmination of years of work, when the final result is unboxed. This could be the most important thing he ever does. "Yes, please!" He's grinning, but inside he is coiled like a spring.

"Zero point eight five, plus minus point zero four." A broad smile now breaks across Paula's face. "That's over three sigma!" Suddenly everyone is smiling. They have just found evidence of a brand-new force of nature. Mitesh slumps back into his chair, stunned. "Congratulations, everyone," he says.

February 25, 2021, Fermilab, Illinois

In an office at Fermilab, a sealed envelope has rested safe in a locked drawer for three years. Inside, scrawled in Biro on a slip of folded paper, is an eight-digit number, the secret frequency needed to decode the results of one of the most long-awaited experiments in science.

A few hundred meters away in the main experimental hall is the Muon g – 2 experiment, a hulking blue magnetic ring, fifty feet in diameter. For two years, subatomic particles called muons flashed around it at close to the speed of light, their decays recorded by a bank of detectors arrayed around its center. But now the machine is quiet, the magnet powered down, the precious data recorded and analyzed. All that's needed is the frequency.

Only Greg Bock and Joe Lykken have ever seen the eight-digit number. They are the frequency's neutral, disinterested (although not uninterested) guardians, protecting the experimental team from the temptation to consciously or subconsciously massage their results. There is no room for bias when the outcome could herald a new age in physics. Like a cipher, the frequency blinds the team to their own results, that is, until today.

After years of work perfecting their analysis, Chris Polly, co-spokesperson of the experiment, finally retrieved the envelope earlier that day. Sitting alone in the control room later that afternoon, pale winter sunlight shining through the windows and with 170 of his collaborators watching on tenterhooks over the internet, Polly tears open the envelope and holds the frequency up to his webcam for all to see. A moment later, when the result is unscrambled, his colleagues erupt in applause.

June 5, 2019, Blois, France

At the start of summer 2019, I took part in a physics conference held in the Château de Blois, an opulent seventeenth-century French castle perched on a hillside overlooking the river Loire. Particle physics is rarely a glamorous business; I spent the three years of my PhD studies in a windowless office, which, rather distressingly, was located directly beneath the frequently leaky first-floor men's bathroom (water dripping unexpectedly onto my head and the smell of bleach still give me flashbacks). But the Blois conference was definitely an experience to be savored. In particular, the lunches.

This being France, lunch was an extended, four-course affair: pâtés, salads, chunks of meat swimming in rich creamy sauces, not to mention the desserts, all amply complemented by wine. The upshot of which was that once the afternoon talks began, at least half the delegates were too sozzled to stay awake, let alone follow a physics presentation. The first slot of the afternoon was a particularly tough gig. Even if the speakers had abstained at lunch themselves, they still had to contend with a smattering of delegates gently snoozing in the high-ceilinged ballroom where the sessions were held.

The unenviable task of waking the audience from their boozy stupor on the Wednesday afternoon fell to Cristian Rusu, a researcher from Japan's National Astronomical Observatory. He was there to present the latest results from something called H0LiCOW, which I rightly presumed was a tortured astronomical acronym, rather than anything to do with actual farm animals. The conference deliberately brought together particle physicists (like me), who study the smallest building blocks of nature, and cosmologists, who study the universe at the vastest scales imaginable, with the aim of encouraging collaboration between the two fundamental sciences. As a result, about half of the talks at the conference were well outside my usual comfort zone, but I was enjoying getting to hear about research that didn't just involve failing to find any signs of new particles at the Large Hadron Collider (LHC), an all-too-familiar refrain from my own kind.

It quickly became clear that Rusu's talk was arousing a lot more interest than was usual for the post-lunch session. He had begun by solemnly requesting that everyone keep the results he was about to share within the four walls of the conference hall; they were yet to be published, and he didn't want to scoop the official announcement. When I navigated to Rusu's slides on the conference website, I discovered that the key result he was presenting had been blacked out. Something was clearly afoot.

H0LiCOW, I learned, is a collaboration of cosmologists who use images taken by the Hubble Space Telescope to measure how fast the universe is expanding. Around 13.8 billion years ago, the universe burst into existence from a tiny point, far smaller than an atom, growing in size unimaginably quickly in a fearsome explosion of space, time, energy, and matter: the big bang. The universe has continued expanding ever since, which means that when astronomers look into the deepest reaches of space, they see that almost every galaxy in the night sky is rushing away from us.

This effect was first discovered by the American astronomer Edwin Hubble in the late 1920s, who found that the farther away a galaxy was from Earth, the faster it was retreating. In fact, he discovered that there was a straightforward "linear" relationship between how far away a galaxy was and how fast it was moving—double the distance to a galaxy, and it moved away twice as fast. As a result, to figure out how fast a galaxy ought to be moving away from Earth, you simply need to multiply its distance from Earth by a number called the Hubble constant, which in essence tells you the rate that the universe is expanding. The bigger the Hubble constant, the faster space is stretching.

Measuring the Hubble constant ever more precisely is one of the great missions of cosmology. Its value not only tells us about how space is expanding but also helps cosmologists unravel the history of the universe and predict its eventual fate. There are, broadly speaking, two ways to do this. One set of methods looks out into the "local universe"* and measures the distances and speeds of things that emit light, like stars, supernovas, and galaxies. The other approach uses the oldest light in the universe, the faded afterglow of the big bang itself, known as the cosmic microwave background. These so-called early universe measurements calculate the composition of the infant cosmos from patterns in the cosmic microwave background and then extrapolate forwards in time to estimate how fast space ought to be expanding today. In principle, since these two methods are measuring the same thing, they ought to agree.

Unbeknownst to me as I sat in that conference hall in Blois, a crisis in cosmology had slowly been brewing for a number of years. It hinged on the value of the Hubble constant. The local and early universe values, it turned out, had been slowly diverging from each other as each set of measurements got

* "Local" here is a relative term; it includes anything up to around a billion light-years away.

ever more precise. However, the two sets of measurements hadn't yet reached the precision necessary to be sure that it was a real effect and not just a bit of bad luck—that is, until Rusu presented his results.

H0LiCOW's aim was to measure the Hubble constant using the most energetic and violent objects in the cosmos, quasars. Quasars are found in the hearts of distant galaxies and are thought to be caused by supermassive black holes, with masses of millions or billions of suns, devouring huge clouds of dust, gas, and stars before blasting out terrifying beams of matter and radiation with a power that can be thousands of times greater than all the stars in the Milky Way. Their terrific brightness means that they can be seen by telescopes even at enormous distances, which makes them an excellent way to track the cosmos's expansion.

The reason for the cloak-and-dagger atmosphere around Rusu's results soon became clear. H0LiCOW's measurement of the Hubble constant had finally caused the tension between the local and the early universe measurements to grow past the critical statistical threshold known as five sigma. Without getting into the statistics in detail—we'll save that for later—in essence this meant that there was now less than a 1 in 3.5 million chance that the two sets of measurements disagreed with each other due to a random statistical wobble.

In other words, something very strange indeed was going on in the heavens.

This is what in physics is known as an anomaly, an unexpected result that seems to contradict our accepted ideas of how the universe works. A growing list of such anomalies has been cropping up in a range of experiments over the past decade. Some have appeared in particle collider experiments like the one I work on at CERN, the European Organization for Nuclear Research. Others have turned up in detectors that search for particles raining down from space, yet others from studying the universe as a whole. This particular anomaly—known in the cosmo biz as the Hubble tension—is the big one. It implies that our basic model of the universe is wrong, challenging the accepted history of the universe and potentially implying the existence of cosmic forces never previously imagined. The fate of the universe itself is, quite literally, in the balance.

The controversy surrounding the Hubble tension became all too clear once Rusu had finished giving his talk. Cosmologists around the room leaped up to ask questions, some attempting to unpick H0LiCOW's methods, others speculating about what it could all mean for our understanding of the universe. Some members of both the early and the local universe camps have been keen to blame the other side for making a mistake in their analysis, each convinced of the rightness of their own favored method and suspicious that the other side might have blundered. That our long-standing and, so far at least, stunningly successful cosmological model may be wrong, and that the universe may be even stranger than anyone supposed, was too much for some in the room that day to countenance.

Several years on from the conference in that French château, the Hubble tension has grown only stronger and more mysterious. Both sets of measurements have been pored over and repeated with ever greater care, and no one has managed to find a mundane explanation for this strange divergence.

Meanwhile, in particle physics, a series of anomalies has also been growing ever more significant, leading many to believe that we are on the brink of something big. Particles with unbelievable energies have been spotted bursting from beneath the Antarctic ice while hidden forces seem to be tugging on the basic building blocks of matter. From the vast subterranean caverns of the Large Hadron Collider to a balloon floating high above the frozen ice sheets of the South Pole, scientists are uncovering a catalog of weird phenomena that can't be explained by our long-established theories of the universe.

There's been a lot of talk lately about a crisis in physics. Despite being confronted with some of the deepest mysteries we've ever faced in science—from the nature of the invisible dark matter that dominates the universe, to the strange fact that the particles that make up our world should have been annihilated during the big bang—every attempt to resolve them has crashed and burned as experiment after experiment has failed to offer any clues on where to go next. It's gotten so bad that some have even started to fear that particle physics could be reaching a dead-end and that these mysteries will remain unsolved—perhaps forever.

So, the stakes (for physicists at least) could not be higher. These anomalies could be the answers to all our prayers, lifting the veil on nature's best-kept secrets and leading to a revolutionary new scientific age.

But they also bring danger. As Carl Sagan once said, extraordinary claims require extraordinary evidence, and there are good reasons to question whether these results are really signs of something wondrous and new or mere statistical flukes, theoretical confusion, or, worst of all for us practitioners, a glitch in an experiment. Could we be so desperate to see something new that we risk seeing effects where there are none? Could we be chasing ghosts? Or, after decades of fruitless searching, could we finally be catching glimpses of a profound new view of the fundamental nature of reality?

These are the questions at the heart of this book. How they get answered could transform how we think about the cosmos.

The search for answers will take us on a journey across the globe, from a mysterious underground vault on the outskirts of Chicago to telescopes perched high above the Atacama Desert. Along the way we'll meet the men and women on the hunt for scientific glory, whose careers and reputations are staked on these uncertain results. As a physicist working on one of these anomalies myself, I'll share my own firsthand account of the drama unfolding at CERN's Large Hadron Collider. We'll also meet the hardened skeptics who dismiss the anomalies as mere cock-ups or wishful thinking, and the theorists following the experimental bread crumb trail as they attempt to figure out what all this could mean for the future of physics—from new fundamental forces to a wholesale revision of the big bang theory.

Along the way, I hope to share some of the excitement of working at the cutting edge of modern physics as we crisscross the globe, and journey through time and space, from the farthest galaxy to the tiniest constituents of matter.

Our story begins, fittingly, at the beginning.

Chapter 1

The Cosmic Story

*The origins of the universe, the nature of matter,
and what we still don't understand*

Despite what you might have been told, the universe
didn't begin, not really.

It's tempting to picture the embryonic universe,
before the big bang, as a seed, pregnant with all the possibility of the cosmos to come, ready to germinate. But as far as we
understand, the universe was already growing when it began.
Which means it wasn't a beginning at all, but rather the start of
a new phase of its existence. But for simplicity's sake, and since
all stories need a beginning, let's say that our universe began
with growth.

But this was not the generative growth of a germinating seed.
This was a lacerating, annihilating growth, one that destroys,
tears, and rends. It was the growth of space itself.

At the start of everything, the cosmos grew at a truly terrifying rate, expanding exponentially. Space begat space begat
space. Reality was stripped of all form by space's expansion,
becoming a tenuous, near-featureless void. If you could have sat
on an atom—though there were no atoms, not yet—and gazed
out at the rapidly expanding universe, you would have seen
nothing, only a freezing, suffocating darkness. The remorseless
stretching of space would have carried all other objects far out

of sight, faster than the speed of light, beyond an unreachable horizon. Paradoxically, as space expanded, the world visible to any would-be observer shrank to a tiny, inescapable nutshell, binding every speck within its own isolated sphere of nothingness, separated from all other such specks by an ever-growing void.

This age of extreme expansion has been given a rather mundane name: "inflation." How long inflation went on we do not know. It could have been a mere instant, or an eternity. In all probability, we will never know. Nor are we likely to discover what, if anything, came before it took hold of the cosmos. Any traces of earlier epochs have been erased by the growth of space.

But we do know that at some point inflation ended, and as it ended, the universe roared into being. As the force that drove inflation dwindled, its power birthed innumerable particles, filling the gasping void with a searing, subatomic fire, the primordial plasma from which all objects in the cosmos would later form.

From darkness, light. Within a trillionth of a second, the forces of nature came into existence and with them the first and greatest blaze of light in cosmic history. Though inflation had ended, space was still expanding, albeit at a less ferocious rate. As the cosmos continued to grow, the primordial fire began to cool. Elementary particles, the building blocks of everything yet to come, fused to form the hearts of the first atoms. Over the next few minutes, the nuclei of the first chemical elements were forged: hydrogen, helium, and lithium.

For around 380,000 years, the infant universe was filled with an incandescent fire, cooling slowly as space continued to swell. Then, at a critical moment, the temperature dropped low enough to allow electrons to wrap themselves tightly about wandering atomic nuclei, forming the first whole atoms. At this moment, the primordial flame went out and darkness fell throughout the cosmos.

A long dark age began. For hundreds of millions of years, the universe was filled by a warm hydrogen fog. But in the darkness, the first seeds of structure were gathering. Slowly at first, then faster and faster, hydrogen and helium were drawn under gravity into ever-thicker clouds. As these clouds grew dense, they also grew hot. At last, somewhere in the unsearchable reaches of space, the pressure at the center of one cloud became so intense that it ignited a nuclear fire, the spark of the first star. Light flooded back into the universe and the dark age ended with a cosmic dawn.

Soon, the universe was filled with stars, bright points of light in the slowly expanding dark. Gathered in the first galaxies, they forged chemical elements never seen before. Carbon, oxygen, nitrogen, iron, and many more were scattered through space as these first stars died in blazing supernovas. From their glowing embers, new generations of stars were born, and yet more elements were forged, a continuous cycle of birth, death, and rebirth.

Around nine billion years after the cosmic dawn, in an outer spiral arm of one of the hundreds of billions of galaxies now spread through space, a new star flickered into light. About it circled a fleet of worlds, and on one rocky sphere inert matter somehow came to life. Under the guidance of dumb physical forces, atoms arranged themselves into ever more complex organisms that grew, multiplied, and evolved. For billions of years, the pale blue orb teemed with living things, species emerged and perished, and the dance of evolution went on. Until, at last, humankind stood on the surface of the Earth, gazed up at the stars, and wondered.

—

This story of the cosmos, its origins, and its history has been gradually pieced together over centuries by the collective effort of thousands of people. The fact that modern science allows us to speak with such confidence about events so remote in time and space that they defy imagination is surely one of the greatest achievements of our species. Particularly when you consider that this story was assembled by peering and prodding at the universe from a rather restricted vantage point on the surface of an insignificant little rock hurtling through space.

The evidence that underpins this cosmic story comes from two sources. The first is the sky. By looking up into the heavens, first with our eyes and later using clever instruments made out of the physical matter of the Earth, we have learned about stars, planets, gravity, the expansion of space, and the fireball of the big bang. However, the problem with stuff in the sky is that almost all of it is inconveniently far away. The farthest a human being has traveled is around the Moon, a mere 400,171 kilometers into space, which is an imperceptible shuffle to the right in cosmic terms. We've done a little better using remote space probes. By early 2023, *Voyager 1*, launched by NASA in 1977, had traveled an impressive-sounding 23.8 billion kilometers from Earth, taking the plucky little probe beyond the edge of our solar system and into interstellar space. However, 23.8 billion kilometers is absolute peanuts to the cosmos, a trifling 0.000000000001 percent of the distance to the edge of the observable universe.

The problem, as the novelist Douglas Adams drily put it, is that "space is big." Its bigness makes it impossible to go out and sample a bit of star, or pop over to a black hole and peer over the edge. Instead, we have to make do with inferring the properties of the objects we see in the sky from the signals we passively receive from above. This would have been extremely limiting had it been the only approach we had to making sense of the universe at large. Fortunately, there is a second powerful way to study the cosmos—by probing the physical matter we find here on Earth.

Our modern understanding of matter—the stuff from which we and everything around us are made—ultimately comes from picking up bits of it from our surroundings and doing some rather unpleasant things to them. By means of smashing, boiling, melting, burning, dissolving in acid, electrocuting, hurtling through particle accelerators, and many other torturous practices, humans have slowly discerned that all objects in the world around us are made up of the same basic building blocks: elementary particles. Somewhat miraculously given the enormous variety of things that exist in the world, there appear to be only a tiny number of different types of these particles. You are made from just three—electrons and two types of quarks (more about those soon)—held together by some handy quantum mechanical forces that stop all your bits from falling off.

Over the past century or so we have developed an eerily successful understanding of matter based on experiments performed here on Earth. Using this knowledge combined with the crucial assumption that the stuff we find in our immediate surroundings is the same as the stuff that makes up the wider cosmos, we have extended our intellectual reach far beyond our planet to the most distant stars. For instance, it was discovered in the nineteenth century that when different chemicals are heated in a flame, they give off characteristic colors of light, making it possible to tell two different gases apart just based on what colors they emit. When astronomers break starlight apart, they find it contains the same characteristic colors that are emitted by chemicals here on Earth, allowing them to figure out what the remotest star is made from without the hassle of traveling thousands of light-years to take a sample.

More recently, experiments conducted at the Large Hadron Collider, humankind's largest and most powerful scientific instrument, a twenty-seven-kilometer ring of superconducting magnets that spans two countries and is operated by a global scientific community, have probed how matter behaved under the incredibly extreme conditions that existed a trillionth of a second after the universe came into existence. Knowledge gained from experiments like this makes it possible to pull off the remarkable trick of inferring what the universe was like long before human beings started blundering about here on Earth, and hundreds of thousands of years before even the first atoms formed.

These, then, are the two pillars on which our cosmic story rests: observing the heavens and experimenting on matter. With these two powerful tools we have unraveled much of the history and nature of the universe. We know with some confidence that the universe came into existence 13.8 billion years ago in the big bang. We know that it has carried on expanding ever since and that today that expansion is accelerating. We've discovered how the chemical elements were forged, traced the life cycles of stars, and uncovered the deep laws that govern reality. The cosmic story is full of wonder, beauty, and majesty. It is a testament to human creativity, ingenuity, and curiosity.

And yet it is a story that is far from complete. There are several yawning gaps in the cosmic narrative, many of which appear when our understanding of the universe at large collides with what we know about the microscopic world of atoms and particles.

A Tale of Two Theories

Our understanding of the very large and the very small is encapsulated in two theories, which between them describe more or less everything we know about the fundamental nature of the universe. These two theories are among the greatest intellectual achievements of the human race, dazzlingly successful in describing the world around us, profound, and in places ineffably beautiful. However, since they were cooked up by physicists, who have a less than inspiring track record when it comes to naming things, these two spectacular theories are referred to, rather prosaically, as standard models.

The first, and more fundamental of the two, describes the tiniest building blocks of matter: the standard model of particle physics. Arguably the closest we have to a "theory of everything"—a set of equations that might one day explain all phenomena that have ever been observed in the universe—it describes the ingredients that compose atoms, three of the four known forces that act on matter (gravity, so far at least, isn't included), and a bunch of other elementary particles, with names such as "neutrinos" and the "Higgs boson." The standard model is without doubt the most successful scientific theory ever written down, simultaneously accounting for the structure of matter, how stars shine, and why things have mass. It has passed pretty much every experimental test we have ever thrown at it, often with unbelievable (and sometimes maddening) accuracy.

However, rather strangely perhaps considering its name, the standard model of particle physics isn't really a theory of particles at all. Instead, it describes the world around us as being made of rather more mysterious and nebulous objects known as quantum fields. Particles are fairly easy to picture, conjuring images of little spheres whizzing about in space. Quantum fields are far harder to get our heads around.

Even if we might struggle to picture them, we have all felt the physical reality of a quantum field. I imagine that, given you are reading this book and therefore have more than a passing interest in physics, you will at some point in your life have played with magnets. Magnets, I think, are basically as close as you can get to real everyday magic. I keep a small box in my desk drawer containing around a dozen little gold-colored magnetic spheres, each about the size of a pea. I have lost many a happy half hour fiddling with them, snapping them together into long chains, pulling them apart, or levitating one off the desk using another. But what I find most captivating is taking one in each hand and forcing their two like poles together until they start to push back. If you do this, you will feel the undeniable presence of a physical *thing* pushing back from between the magnets. Look as hard as you like and you won't see anything in that gap, but there is absolutely no denying that there is something there.

The thing that's there is a quantum field, in this case the magnetic field. It may be invisible, and you may not be aware of it except when you're playing with magnets, but nevertheless it is ever present. And it is not simply the case that magnets generate a magnetic field close to their poles. In truth, the magnetic field permeates *the entire universe*, connecting my little magnets to the magnetic dynamo of the Earth, the Sun, and even the most distant galaxy. There is only one magnetic field, and it's everywhere.

The "quantum" bit of a quantum field comes into play when we try to understand what a particle is. If quantum fields are like invisible fluids filling all of space, particles can be thought of as little ripples in these fluids. The reason you are able to read these words is that a torrent of uncountable particles of light known as photons are bouncing off the page and smashing headlong into your retinas. Each photon is a tiny undulation in something called the electromagnetic field (of which the magnetic field is just one aspect). What we think of as a particle, therefore, is simply the smallest possible amount of wobble that can travel through a quantum field.

It may seem counterintuitive to think that it is the invisible influence of fields that governs the universe. But in fact, it's even stranger than that. According to the standard model, fields explain not just light and photons but even the particles that make up the apparently solid world around us. Atoms are ultimately made of subatomic particles—electrons and quarks—which aren't tiny hard spheres, as we've been trained to imagine, but instead little ripples in underlying fields. Electrons are little ripples in something called the electron field, while quarks are tiny wobbles in the quark fields.

It is these fields, not particles, that are the ultimate constituents of our universe. Each of us is made of fields. Deep down, we are all vibrations in the same invisible oceans.

That said, particles are the things we actually see in our experiments, and they remain a very useful way of thinking about the world. Throughout this book I'll discuss particles again and again; I am a particle physicist after all. And so, while the standard model of particle physics tells us that I could just as well refer to quarks as "quantized ripples in the quark field," in the interest of brevity I'll stick with old-fashioned "quarks."

Quantum fields are described using an incredibly powerful and predictive theoretical framework known, unsurprisingly, as quantum field theory. Quantum field theory is the language in which the standard model of particle physics is written, in the same way that English is the language of Shakespeare.

And like the English language it can be used to do some pretty extraordinary things.

To get a sense of just how powerful quantum field theory is, consider, for a moment, the first fundamental particle to be discovered: the electron. As well as being negatively charged, electrons behave like tiny bar magnets, with a north and a south pole. Using the known quantum fields in the standard model, it is possible to calculate how strong the electron's little bar magnet ought to be to a frankly ludicrous level of precision. Here it is:

$$0.00115965218161$$

Don't worry about what this number means too much or what the units are. The important thing to appreciate is that this number is calculated to fourteen decimal places! Now, if you do an extremely clever and careful experiment, it is possible to measure the electron's magnetism to an equally silly level of precision, giving:

$$0.00115965218059$$

You will notice that these two numbers are essentially identical. In fact, the difference between them is a mere 0.0000000000010 or roughly one part in a trillion. What I have just shown you is the best prediction anywhere in science. Ask a biologist or a chemist to show you anything as good as this and I guarantee they will shuffle off looking dejected and probably grumbling about how smug and superior physicists can be.

The take-home message from all of this is that the standard

model of particle physics is *right*. I mean, it just has to be. There is no way you get a prediction this precise by accident.

And yet it is also almost certainly seriously incomplete. And what spoils the party is none other than our second great theory of the universe: the standard cosmological model.

The standard cosmological model covers the opposite end of the scale, zooming out to view the entire universe on the largest stage imaginable. It is the story of the cosmos, its birth, evolution, and eventual fate. Like the standard model of particle physics, the standard cosmological model is based on another, deeper theory. In this case, Einstein's masterpiece, the revolutionary theory of space, time, and gravity known as general relativity.

When Einstein revealed general relativity to the world in 1915, it completely transformed the way we think about the nature of space and time, and in particular the force of gravity. At its core is the profound insight that gravity isn't really a force at all, but a consequence of the way matter and energy bend, warp, and stretch space and time. According to general relativity, space and time aren't merely coordinates that tell you where and when you are—say, having a nice cup of tea at 4:00 p.m. in your favorite armchair—but a dynamic, interwoven fabric. In other words, space and time are a physical *thing*. While Isaac Newton argued that the Moon orbits the Earth thanks to Earth's gravity reaching out across the void of space and tugging on the Moon, Einstein argued that there's no such thing as gravity. Instead, the mass of the Earth literally bends the physical, stretchy fabric of space and time around it, like a bowling ball resting on a trampoline, forcing the Moon to follow a bent path in this curved space.

But general relativity can do far more than just explain the orbits of heavenly bodies; it can also be used to describe the behavior of space at the grandest possible scales—the scale of the entire universe. The advent of general relativity allowed scientists to write down equations describing the cosmos as a whole, giving birth to the modern science of cosmology.

The power of general relativity is that it tells you how the elastic fabric of space and time responds to the presence of different types of matter and energy. For instance, ordinary matter—the stuff that makes up galaxies, stars, and you and me—tends to cause space to contract. Meanwhile, there are altogether weirder types of energy that can do the opposite, acting as a form of repulsive antigravity that makes the universe expand. That means if you want to cook up a model for the cosmos, all you need to do is decide on what ingredients you'd like your universe to contain—for instance, a dash of ordinary matter, a sprinkling of radiation, and perhaps some repulsive energy—and general relativity will tell you how your model universe will evolve.

As we saw briefly at the start of this chapter, the standard cosmological model begins with a universe dominated by a powerful form of repulsive energy. This repulsive energy caused an incredibly rapid period of expansion, which cosmologists call inflation. In an extremely short time—around ten billionths of a trillionth of a trillionth of a second—the universe exploded in size by a factor of ten trillion trillion. Now, I'll readily admit that these are some pretty insane numbers. I mean, what on earth does ten billionths of a trillionth of a trillionth of a second feel like? I could just as well have told you that the universe expanded by a squillion bazillion times in one iota of a yoctosecond. The key message is that the universe got bigger *really* fast.

The next thing that happened was that inflation ran out of

puff and the repulsive energy that drove it converted into a mixture of matter and radiation. This created a fiery inferno of particles that rapidly filled the nascent universe and led to what cosmologists call the hot big bang. It is from this maelstrom that all the objects in the universe today would ultimately form. From this point onward, the history of the universe was dictated by precisely what kinds of matter and energy emerged from the death of inflation and in what quantities.

And this is where the trouble starts. Because, as far as cosmologists can tell, about 95 percent of this stuff seems to be missing.

Through careful studies of the light left over from the big bang, cosmologists estimate that the stuff that makes up you and me, the planets, the stars, and everything we can see is just a small fraction of what's really out there. The cosmos, it seems, is dominated by two mysterious substances: dark matter and dark energy. The first firm evidence for dark matter was discovered in the 1970s by the American astronomer Vera Rubin, who found that stars orbiting our closest galactic neighbors, such as Andromeda, appeared to be moving far too fast for the gravity provided by all the visible stuff—gas, dust, and stars—to hold them on course. By rights, the stars should have been flying free from their galaxies, skidding into intergalactic space like a car that had taken a corner too fast. And yet somehow the galaxies stayed together.

The solution was to propose the existence of a vast amount of invisible stuff whose gravitational pull keeps the stars attached to their galaxies: dark matter. Today, the evidence for dark matter is overwhelming, with all but a small minority of astronomers, astrophysicists, and cosmologists convinced of its existence. Our best estimates suggest that there is around five times more dark matter in the universe than visible matter.

And by visible matter I mean literally everything we can see in the night sky, every galaxy, every star, every cloud of gas, and every speck of dust. The problem is, we haven't the foggiest clue what dark matter actually *is*.

What we can say with a high degree of certainty is that dark matter isn't made of atoms, or indeed any other particle we've discovered in experiments. The standard model of particle physics describes seventeen different elementary particles, but none of them has the right properties to be dark matter. It is hoped that we will eventually be able to detect dark matter particles directly as they drift through the Earth, or perhaps even create them in particle colliders. But so far at least, we haven't had any luck.

Even more enigmatic than dark matter is dark energy. In the mid-1990s, careful observations of exploding stars revealed that the expansion of the universe appears to be speeding up, overturning the established view that it should be slowing down due to gravity's tendency to pull everything back together. The most popular explanation for this accelerating expansion is a strange form of energy that permeates all of space, generating a repulsive force that overwhelms gravity at sufficiently large distances and drives the universe to expand faster and faster. Cosmologists call this unknown thing dark energy. Dark energy is thought to account for around 68 percent of the total energy content of the universe, with dark matter making up around 27 percent and ordinary matter (that is, you, me, and everything in the night sky), a measly 5 percent.

Again, we have no real clue of what dark energy is. Some kind of force field perhaps? Or the energy intrinsic to space itself? Or perhaps it's something we have yet to imagine. One take-home message from all of this is that when you hear the word "dark" being used by physicists, you should get very suspicious because it generally means we don't know what we're talking about.

If you turn to the standard model of particle physics for an explanation of dark energy, things go badly wrong. *Very* badly wrong. Attempting to estimate how strong dark energy ought to be based on what we know about quantum fields results in what has been called the worst prediction in science, giving a number that's 10^{120} (that's 10 with 120 zeros after it) times larger than what we observe in the cosmos. If dark energy were anywhere near as strong as particle physics suggests, then the whole universe would be ripped apart in an instant.

Now, you might think that having no explanation for 95 percent of the universe is a pretty big problem, and you'd be right. If you were a cartographer and 95 percent of your maps had "Here be dragons," scrawled all over them you might have a hard time taking yourself very seriously. But the cosmic story has more than just gaps; in some places it contains outright paradoxes. One of the most troubling of these concerns how matter came to exist in the first place. Particle physics tells us that every matter particle has a mirror image, identical in every way but with the opposite charge. These mirror particles are called antiparticles. The good old negatively charged electron that we find whizzing around every atom has an antiparticle with positive charge known as the positron. Likewise, for each of the quarks that make atomic nuclei, there is an equal and opposite antiquark. Now, according to our current best theory of particle physics, whenever you make a particle, say by bashing some atoms together very hard, you must also make an antiparticle. Similarly, when a particle meets its antiparticle, the two will annihilate each other, disappearing from existence in a flash of radiation. This state of affairs has been confirmed by every experiment we have ever conducted. However, when we apply this understanding of particles and antiparticles to the very first moments of the universe, things go disastrously wrong.

In the first millionth of a second after the big bang, the universe was so hot that particles and antiparticles were continuously being created out of energy, emerging from the seething primordial plasma before annihilating again. Each time a particle was created, so was an antiparticle, and whenever an antiparticle was annihilated, so was a particle. Equality between matter and antimatter reigned.

However, around a millionth of a second into cosmic history, the universe expanded and cooled to the point where there was no longer enough heat in the primordial fireball to make new particles and antiparticles, and an event known as the great annihilation took place. All the particles and antiparticles created in the first millionth of a second rapidly destroyed each other in a fearsome blast of radiation, leaving a universe made of naught but light.

In other words, particle physics tells us that the universe should contain nothing: no stars, no planets, no you or me.

And yet, perplexingly, here we are. Our very existence, and that of the material universe, is an enormous challenge to our understanding of particle physics. Again, what we see in the universe at large—that there is anything to see at all and that we are here to see it—conflicts violently with what we have learned by studying the smallest components of matter. Something deep is clearly missing from our current telling of the cosmic story.

Now, I don't mean to make this sound like a bad thing. For a scientist, ignorance is a kind of tortured bliss. Problems like these mean that there is still more to learn, and the bigger the problems, the better. It is trying to solve these mysteries that makes science worth doing. Wonderful and awe inspiring as what we already know is, it is the process of discovery that really sets the heart aflutter, what Richard Feynman called "the pleasure of finding the thing out." I can think of nothing more depressing than a world where we'd learned all there is to learn, where there was nothing new to add to the cosmic story.[*] The existence of these gaps and paradoxes means that there is a chance to help write parts of this grand narrative, and I can think of no greater privilege than that.

What is rather more troubling is that we have been facing many of these big mysteries for decades, in some cases for more than half a century, and so far, at least, every attempt we have made to find solutions has failed. Speculative new theory after speculative new theory has come a cropper when confronted by cold, hard experimental evidence (or more often, lack thereof). Worse still, some of the solutions to these problems make no testable predictions at all, calling into question whether we should really regard them as scientific.

Things have become so acute in the past decade that a view has gotten about that particle physics in particular might have reached a dead end. And if particle physics were to die, cosmology too would hit the buffers. Dark matter, dark energy, inflation—key ingredients of the cosmic story—would remain forever unexplained, mere words to cover our fundamental ignorance. If this rather depressing take is right, then the quest to understand the universe, its nature, and its origins might be coming up against the limits of the knowable.

---

[*] Okay, that's not really true. I can think of plenty of things that are more depressing than science being finished. A world ruled by a nihilistic death cult, for instance, or one without chocolate digestive biscuits. But you get my drift.

But in the last few years, chinks of light have started to appear in the form of unexplained anomalies. Could they be the clues that allow us to write the next chapter of the cosmic story? That is the question on the mind of every particle physicist and cosmologist right now, and the focus of this book.

Soon enough, we're going to get to the cutting-edge results that are challenging our current theories. But before we do, it's worth taking a short detour to appreciate the part anomalies have played in deepening our understanding of the world. For as we'll see, anomalies played crucial roles in building the twin pillars of modern physics, general relativity and quantum field theory, just as they may help pave the way for our next epoch-defining discoveries.

Chapter 2

The Fall of Planet Vulcan

*How two anomalies revolutionized our
understanding of the cosmos*

On December 31, 1859, the country physician Edmond Lescarbault was at his home in the small French village of Orgères-en-Beauce, about sixty-five miles southwest of Paris, when there was an unexpected knock at the door. Perhaps he had been sitting by the fire writing cards to give to his friends and neighbors the next morning, or simply looking forward to a quiet New Year's Eve meal with his family. He almost certainly wasn't expecting to find himself confronted by the commanding figure of Urbain Le Verrier, director of the Imperial Observatory in Paris and France's foremost astronomer, famed for his role in the discovery of the planet Neptune.

When his duties as a rural doctor allowed, Edmond Lescarbault was an enthusiastic amateur astronomer and had even built his own small observatory next to his house. Just over a week before Le Verrier's unannounced visit, Lescarbault had been leafing through a popular astronomical journal when his eyes alighted on an article describing a recent piece of work by Le Verrier on Mercury, the closest planet to the Sun. Mercury's orbit around the Sun had been puzzling astronomers for decades; the planet seemed to stubbornly refuse to follow the path laid out for it by Newton's law of gravity. For more than a century, astronomers had been embarrassed by the little planet while attempting to observe its transits, rare astronomical events when Mercury passes directly between the Earth and the Sun, appearing as a small black dot against the Sun's brilliant surface. Mercury repeatedly failed to appear when predicted, sometimes causing astronomers to miss transits by hours, or in one extreme case, by more than a day. In Newton's predictable clockwork universe, a planet with such poor timekeeping was extremely disturbing.

Le Verrier had made his name by solving a similar problem with the orbit of the outermost planet, Uranus. Like Mercury, Uranus's orbit didn't conform to the predictions of Newton's theory, an anomaly that Le Verrier had claimed could be explained if its orbit was being perturbed by the gravitational tug of an unseen eighth planet. In a virtuoso calculation, Le Verrier was able to predict precisely where this eighth planet ought to appear in the night sky, and when astronomers looked through their telescopes, they discovered a brand-new planet—Neptune—less than one degree from the point where Le Verrier had said it should be.

Emboldened by his previous triumph, Le Verrier was now hoping to pull off the same feat with the planet Mercury. Earlier in December, he had published a paper suggesting that Mercury's orbit could be accounted for if there existed a hitherto undiscovered planet between Mercury and the Sun. A small world, or perhaps a collection of asteroids, hiding in the dazzling glare of our local star could well have escaped notice, he argued.

On reading the article, Lescarbault's heart began to race. Nine months earlier, on March 26, he had been observing the Sun using his handmade telescope when he had noticed a small dark object on its surface. He had quickly convinced himself that it wasn't a sunspot and tracked the object until it disappeared over the edge of the Sun's disk. While a curious observation, Lescarbault was unsure of exactly what he had seen, and hoping to catch sight of it again, he kept his discovery to himself, that is until he read of Le Verrier's prediction. Thrilled by the prospect of having discovered a ninth planet, Lescarbault fired off a letter to Le Verrier in Paris, telling him that he might have sighted his hypothesized world. What he hadn't anticipated was that his letter would elicit such a strong response that Le Verrier would immediately jump on a train from Paris and would, just a few hours later, be hammering on Lescarbault's front door.

With barely a "how do you do," Le Verrier forced his way into Lescarbault's hallway demanding to inspect his observatory and records while berating the startled doctor for sitting on what could be a momentous discovery for almost nine months. After submitting Lescarbault to a fearsome interrogation (which one account compared to the meeting of a lion and a lamb), he proceeded to examine his homemade equipment and, finding it relatively satisfactory, made some minor corrections to Lescarbault's observations of the mysterious object. Apparently not content with this, he then proceeded to demand character witnesses from Lescarbault's neighbors, presumably to make sure he wasn't being taken for a ride by a confidence trickster. By the end of what appears to have been a rather grueling day for Lescarbault, Le Verrier returned to Paris, assured that his predicted planet had been found. Given its nearness to the furious heat of the Sun, he named it after the Roman god of fire, Vulcan.

Announced to the astronomical community by Le Verrier, Lescarbault's discovery caused a nationwide sensation, with newspapers across France reporting on the remarkable new planet. Lescarbault found himself catapulted from unknown country physician to astronomical superstar and was made *chevalier de la Légion d'honneur* (knight of the Legion of Honor) just a month later. Meanwhile, Le Verrier basked in the glory of a second audacious prediction proved true. By February, news had spread to France's imperial rival Britain, and while there was no shortage of dismay that the French had beaten Old Blighty to yet another planet, the story proved irresistible, with newspapers and public lectures spreading word of the tiny, scorched planet Vulcan.

Le Verrier soon found himself inundated with reports of further observations of the new planet as it crossed the Sun's disk. With each one, he adjusted his calculations of Vulcan's orbit and issued new predictions for when the next transit might be seen. When the predicted transits failed to materialize, he was forced to adjust his orbital model and issue new predictions. Sporadic sightings of Vulcan continued to come in over the next few decades; however, it proved impossible to make accurate predictions of when the planet would show up. As the years rolled by, skepticism began to grow about whether Vulcan really existed at all, although Le Verrier clung to belief in his new world until he died in 1877.

Astronomers would continue to scour the skies for Vulcan into the early years of the twentieth century. It was hoped that the planet might become visible during total eclipses, when the Sun's glare was blocked by the Moon, allowing a small pinprick of reflected light to stand out against the temporarily darkened sky. However, despite attempts to spy Vulcan during the eclipses of 1883, 1887, 1889, 1900, 1901, 1905, and 1908, the planet failed to make an appearance. Particularly conclusive were thorough photographic searches made by the American astronomer Charles Dillon Perrine during the solar eclipses of 1901, 1905, and 1908. By 1908, William Wallace Campbell, director of California's Lick Observatory, was ready to declare that the search for Vulcan had finally been brought "to a close."

But despite Vulcan's demise, the problem of Mercury's anomalous orbit remained. Vulcan or no, there was still something seriously fishy about its motion around the Sun, and if it wasn't caused by a new planet, then what? The solution, when it was eventually found, turned out to be far more profound than anyone could have imagined.

To understand the specific issue with Mercury's orbit, the first thing to know is that planets don't orbit the Sun in perfect circles as Nicolaus Copernicus had argued when he published his Sun-centered model of the universe in 1543. Instead, they orbit in ellipses. That means that at some times of their years they are closer to the Sun than at others. In fact, of all the planets, Mercury's orbit shows the biggest difference between its closest (forty-six million kilometers) and its most distant (seventy million kilometers) points from the Sun. Astronomers call the closest point to the Sun on a planet's orbit the perihelion, which is just a fancy ancient Greek way of saying "closest to the Sun."

The second thing to know about Mercury's orbit is that the elliptical path it follows around the Sun doesn't stay fixed in space; the ellipse itself slowly rotates around the Sun over time, changing its orientation. You can imagine Mercury's orbit as a squashed hula hoop, with the Sun acting like the person swinging the hoop around their waist. The reason that Mercury's orbit gradually changes its alignment is thanks to the gravitational tugs of the other planets in the solar system, particularly from its nearest neighbor, Venus.

The rotation of Mercury's orbit causes the perihelion to gradually move around the Sun over the course of many Mercurian years. It was this phenomenon that astronomers were failing to accurately account for. The perihelion appeared to be advancing faster than predicted by Newton's law of gravity, meaning that Mercury completed an extra orbit once every twelve million Mercurian years[*]—a tiny effect, but enough to explain why astronomers kept missing its transits.

At the very time that the search for Vulcan was reaching its disappointing conclusion, a young Albert Einstein was embarking on a journey that would ultimately lead to his greatest contribution to science: the general theory of relativity. This journey would take Einstein through a labyrinth of abstract mathematics, so mind-meltingly hard that few mathematicians at the time, let alone physicists, were able to understand it. Einstein's objective was nothing less than a revolution in our understanding of gravity, space, and time, and it would take him a full eight years to complete his magnum opus.

Einstein's general theory of relativity promised to overturn Newton's law of gravity, which had ruled the heavens for 250 years. Instead of describing gravity as a force exerted between two massive bodies—say the Sun and Mercury—Einstein's theory said that gravity was an illusion created by the way massive bodies warp the fabric of space and time. The difference was more than just conceptual; it had real-world effects. The Sun has a huge mass, meaning that space-time close to the Sun—that is, around the orbit of Mercury—is highly curved, much more so than it is for the other planets. This means that Mercury would be especially sensitive to the differences between general relativity and Newton's law.

By 1915, Einstein had almost completed work on his theory. He had long been aware of the problem with Mercury,

[*] A year on Mercury corresponds to roughly 116 days on Earth.

and although it hadn't been his motivation for embarking on general relativity, he realized that if he could account for its weird orbit, it would give a huge boost to the chances of his radical new theory being accepted. By the early twentieth century, painstaking observations by astronomers had resulted in a precise measurement of how much faster Mercury's perihelion was moving around the Sun than predicted by Newton's theory, resulting in a figure of 45 ± 5 seconds of arc per century. "What's a second of arc?" I hear you cry. It's a measure of how far apart two points appear on the sky, but the key thing is to note that it has a value (45) and an uncertainty (5).*

In 1913, Einstein had tried, unsuccessfully, to calculate the advance of Mercury's perihelion, getting a figure of 18 seconds of arc per century—disappointingly far from the measured value of 45. However, two years on, Einstein had carried out a major overhaul of his theory, and when he tried the calculation again in November 1915, to his delight he got an answer in almost eerie agreement with the measured value: 43 seconds of arc. Einstein was so overcome with excitement that he suffered heart palpitations and had to lie down to recover. On November 15 he wrote enthusiastically to his friend Heinrich Zangger,

---

* Here's the longer explanation: A second of arc is a unit used by astronomers to measure the angle between two objects on the sky. A circle contains 360 degrees, and so if you imagine looking at a point on the sky, say a star, through a telescope and then rotating your telescope by 360 degrees, you'll trace out an arc that goes all the way around the sky and back to the same star (obviously you'll end up looking at the ground for quite a lot of time, but imagine you're floating freely in space so there's no annoying planet Earth to block your view). You can divide that circle into 360 degrees. A minute of arc is then $\frac{1}{60}$ of 1 degree, while a second of arc is $\frac{1}{60}$ of a minute of arc. That means there are 1,296,000 seconds of arc in a full circle around the sky.

"I have now derived the up to this point unexplained anomalies in the motion of the planets from the theory. Imagine my good fortune!"

Just three days later, on November 18, Einstein stood up at the prestigious Prussian Academy of Sciences to give a lecture that would change science forever. In it, he outlined, for the first time, his general theory of relativity and, what's more, showed how it had solved the centuries-old puzzle of the planet Mercury. The scientific world was stunned. Einstein's rival David Hilbert was astonished, writing generously to Einstein, "Congratulations on conquering the perihelion motion."

Einstein had solved the Mercury anomaly. In doing so, he ensured that his theory got the attention it deserved.

A revolution in our understanding of the cosmos began that day. Not only had the planet Vulcan been destroyed once and for all, but Einstein's theory would ultimately overturn Newton's theory of gravity, accepted for centuries as the law that governed the universe. The advent of general relativity would lead to the modern discipline of cosmology, the big bang theory, black holes, and much more besides. Today, it is one of the two great pillars of modern physics, arguably the most profound and beautiful theory ever discovered.

What can we learn from the story of Vulcan? Well, perhaps the most obvious lesson is that paying careful attention to small anomalies can lead to major breakthroughs in how we understand the world around us. Of course, we have to be a bit careful here; Einstein didn't start work on general relativity because of the Mercury anomaly. But it did provide the crucial supporting evidence at the moment he felt ready to let his new theory out into the wild. Without its early success in explaining Mercury's orbit, general relativity might have taken far longer to be accepted as the new theory of the cosmos.

It's also interesting to think about the early attempts to explain the anomaly. Le Verrier's invention of a new planet made perfect sense at the time. A very similar problem with Uranus had been solved successfully this way, and crucially, adding an extra planet didn't overturn the fundamental theory of planetary orbits. When you have an established theory that has proven stonkingly successful for centuries, as Newton's theory had, it's far more likely that an anomaly is either down to a mistake with the measurement itself or, failing that, some other cause that lives happily within the existing paradigm. There was nothing revolutionary about conjuring the existence of a planet; Vulcan was just a bolt-on to the existing model of the solar system, one that chugged along nicely according to Newtonian physics. Overturning a whole worldview, as Einstein had, is usually the last resort, because whatever replaces it must not only explain the anomaly but also replicate all the many successes of the previous paradigm without causing a mess somewhere else, for instance by screwing up the orbit of a well-behaved planet like Earth.

The other clear lesson from the Vulcan saga is the absolutely critical importance of precise measurements in flushing out places where our best theories break down. In the wake of his triumphant explanation of Mercury's orbit, Einstein gained a new appreciation for "the pedantic precision of astronomy that I secretly used to make fun of in the past." The unglamorous work of measuring some quantity or other to an increasing number of decimal places can seem like a nerdy obsession. It rarely gets acknowledged in the popular accounts of major breakthroughs. We tend to fixate on lone theoretical geniuses who, we like to imagine, conjure deep truths about the cosmos by the sheer force of their intellect. Experimenters or observers often fade into the background, treated as mere bean counters or fact-checkers. But even Einstein, the theorist's theorist, who summoned up some of the most profound insights ever revealed in the history of science, realized the essential role of punctiliousness and care in experimental science. Those generations of astronomers who had fastidiously tracked Mercury through the heavens might never have solved the planet's riddle, but without them Einstein would have had no data to test his miraculous new theory against, and we'd all be none the wiser.

One in a Million

The anomaly of Mercury's orbit is far from an isolated case. The history of physics is chock-full of weird results that presaged a transformative discovery. Take the other great pillar of modern physics: quantum field theory. As we saw earlier, modern particle physics tells us that the most basic building blocks of our universe are invisible, all-pervading, fluidlike objects called quantum fields. Particles that make up atoms, like electrons and quarks, are just ripples in these invisible quantum oceans. But theorists didn't just wake up one morning and decide to invent quantum fields; at the roots of quantum field theory lies another beautifully precise measurement, this time made by the experimental physicist extraordinaire Willis Lamb.

While many of his colleagues had spent World War II cooking up new and terrifying ways to make things go boom, Lamb had devoted himself to research on microwaves for use in radar. When the war ended, he realized that his newly acquired microwave expertise would allow him to make measurements of atoms at an unheralded level of precision. In particular,

Lamb was interested in measuring what are sometimes called quantum jumps in atoms of hydrogen, which is the simplest possible atom, made of a single negatively charged electron orbiting a single positively charged proton.

Since the 1910s, it had been known that electrons could only orbit atoms in fixed "energy levels." An imperfect analogy is to think of these levels as like the orbits of planets in the solar system, with the electrons as planets and the atomic nucleus as the Sun. The difference is that while a planet can theoretically orbit our Sun at any distance, in quantum mechanics only specific orbits are allowed, a bit like rungs on a ladder, to mix metaphors. If the same rules applied to our planets, it would mean that a planet could move around Mercury's orbit, or Venus's, or Earth's, but not in between.

While electrons might only be allowed to orbit atoms in these fixed levels, they *can* make sudden jumps from one level to another, either by absorbing or by emitting a particle of light—a photon. To jump to a higher level, an electron must gobble up a photon, giving it the extra energy required for the leap. When it falls to a lower level, it radiates a photon. The photon that's emitted or absorbed carries the difference in energy between the two levels. And since the energy of a photon is related to its wavelength—the shorter its wavelength, the higher its energy—this means that each type of atom has a characteristic set of wavelengths that it emits or absorbs, with each wavelength corresponding to a different quantum jump between different pairs of energy levels.

In 1947, Lamb and Robert Retherford, his student at Columbia University in New York, made a superbly precise measurement of a quantum jump between two specific energy levels in a hydrogen atom. For reasons too obscure to go into here, these levels are labeled S and P. According to the best quantum theory of the time,[*] the S and P levels should have had precisely the same energy. But that isn't what Lamb and Retherford found. Instead, they detected a tiny but undeniable difference in energy between the two levels—a difference of just one part in a million.

Now, what's one part in a million between friends? It might be tempting to dismiss such a tiny anomaly as irrelevant or a simple error. But in that minuscule discrepancy lay the clue to a new quantum realm.

In June 1947, a select group of the world's most talented physicists gathered at the Rams Head Inn on Shelter Island, New York, for the first major conference since the end of World War II. The Shelter Island Conference is now legendary as the place where modern quantum field theory was born, and the undoubted star of the show was Willis Lamb and his one-in-a-million anomaly.

Among the relatively small number of participants in the conference were several of the world's leading quantum theorists including Julian Schwinger, Richard Feynman, and Hans Bethe. For the past few years, they had been struggling to come up with a successful quantum theory of the electromagnetic field—the field responsible for electricity, magnetism, and light—but had found their calculations plagued by a rather disconcerting number: infinity.

According to quantum mechanics, nothing can ever really sit still. Like my younger brother aged five, all quantum objects are constantly fidgeting and jittering. This applies both to particles and to fields themselves. If we think of the electromag-

---

[*] The Dirac equation, for quantum aficionados.

netic field as a sort of fluid, then these quantum jitters are like a gentle shimmering on the surface of a lake.

The trouble was that when theorists tried to calculate how much energy was stored in these quantum mechanical jitters, the answer they kept arriving at was *an infinite amount of energy*. Clearly that was nonsense; their energy couldn't be infinite. But try as they might, they couldn't find a way to get rid of the infinities, which effectively rendered their theories useless.

When Lamb took to the stage at Shelter Island, the quantum theorists present realized they were being given a vital clue that might finally allow them to tame infinity. To understand why, we must first consider a little more carefully what an atom really looks like.

The familiar, easy-to-picture solar system analogy for an atom is, as with so many analogies in physics, rather misleading. Planets go around the Sun in well-defined elliptical orbits. Electrons on the other hand are not little planets but quantum mechanical particles that behave in a wholly counterintuitive way. Since electrons obey the laws of quantum mechanics, it's impossible to say exactly where an electron is at any given moment. This means we can't really think of them as following well-defined orbits as they rotate around the nucleus. In truth, we can only speak confidently about the *probability* of finding an electron at a certain place. So, instead of thinking of an electron as a little planet whizzing merrily around a nucleus, we're better off thinking of it as a hazy cloud. The cloud is thickest where you are most likely to find the electron, and thinner in places where the electron is less likely to turn up.

The *S* and *P* levels in the hydrogen atom might have been predicted to have the same energy, but they were distinguished by their *shape*—the hazy quantum mechanical cloud that the electron adopted when occupying either of the two levels. The S level was spherically symmetric, a bit like a fuzzy ball of cotton wool, densest at the very center and thinning as you move farther away from the atomic core. The P level didn't have the same spherical symmetry: instead, it had two lobe-like structures extending out from the nucleus, like an hourglass, thinning to nothing at all at the very center of the atom.

This difference in shape turned out to be crucial. At the very center of the hydrogen atom sits its nucleus—a single proton. Closest to the proton, the electric field is at its strongest. As a result, the quantum jitters in the electric field are also largest at the center of the atom.

These quantum jitters are like gentle eddies, nudging the electron this way and that as the electromagnetic field wobbles about. The ultimate effect of these eddies is to gently push the electron farther away from the nucleus, changing the shape of the S level so its hazy quantum mechanical cloud is slightly less dense at the center than it would otherwise be. The P level, on the other hand, is less affected as it fades to nothing at the nucleus. This means that the electron in the S level spends more time farther from the nucleus than expected and ultimately causes it to have a slightly higher energy than the P level.

Thanks for bearing with me through that chain of logic there. Here's why it mattered: the assembled theorists saw in Lamb's tiny, one-in-a-million discrepancy the direct signature of the very quantum jitters that had so troubled them in their attempt to construct a quantum theory of the electromagnetic field. With this vital clue, Schwinger, Feynman, Bethe, and others were ultimately able to solve the problem of those nagging infinities.

Once again, from the tiniest of clues, a new view of the universe emerged. Solving the Lamb shift, as it became known, heralded the birth of modern quantum field theory. Particles ceased to be thought of as fundamental objects, and a new paradigm arose wherein every object in the cosmos, from the smallest atom to the largest galaxy, is made entirely from invisible, ethereal quantum fields. Using quantum field theory, physicists were able to make steady progress in understanding the world at the most fundamental level. During the 1960s and 1970s, quantum field theories describing the strong and weak forces (more on these to come)* were discovered, a story that ultimately culminated in the spectacular discovery of the Higgs boson at the Large Hadron Collider in 2012.

So it's important to pay attention to decimal places. Often, a crucial clue is hidden in the last digit. But history has more to teach us than the importance of paying attention to subtle effects. Anomalies can also be dangerous. Woe betide the scientist who fails to heed the warning of anomalies gone wrong: results that promised great leaps forward, dreams of Nobel Prizes, and a place in the scientific pantheon, but ultimately left careers and reputations in tatters.

---

* For now: the strong force sticks quarks together inside protons and neutrons, while the weak force causes, among other things, radioactive decay.

Chapter 3

You Are the Easiest Person to Fool

*Anomalies that led science astray and
how to guard against them*

On March 17, 2014, the world's media gathered at the Harvard Center for Astrophysics for an announcement that would rock the scientific world. A group of scientists working on a telescope at the South Pole declared that they had discovered the "first tremors of the Big Bang," ripples in the fabric of space and time from the very first moments of our universe. Their discovery, they claimed, provided the first direct evidence for inflation, the period of unimaginably rapid expansion that gave birth to the cosmos. And as if that weren't enough, it raised the prospect of probing the relationship between quantum mechanics and gravity, perhaps bringing the dream of an ultimate theory of the universe a step closer. Responding to the result, the theoretical cosmologist Marc Kamionkowski described it as "the greatest discovery of the century." Breathless talk of Nobel Prizes abounded, both for the scientists who had built and operated the BICEP2 telescope that had been used to make the discovery and for the theoretical physicists who had first proposed the theory of inflation.

Thousands of miles away in London, I found myself on BBC Radio, attempting to explain the significance of the breakthrough to a somewhat bemused pensioner from the North of

England named Edna, whom the producers had invited onto the show to make sure I kept things at the right level. I admit I found the whole experience rather humbling, both the enormity of what the BICEP2 team had achieved and my complete inability to convince Edna that it merited her attention. But as it turned out, my science communication failure was small-fry. Within a few weeks, the whole story had unraveled.

It soon became clear that the BICEP2 team had made a catastrophic error. In order to produce their apparently revolutionary discovery, they had taken a key bit of data from a PowerPoint presentation given at a conference by a rival observatory. As other cosmologists dug into their methods, they realized that BICEP2 had grossly misinterpreted the pilfered slide, leading them to underestimate the amount of dust polluting their observations by a factor of four. When this was taken into account, the claimed discovery, arguably the most momentous in living memory, literally turned to dust.

I can't think of a more disastrous example of scientific hubris than the sorry story of BICEP2. It's one thing to make a mistake; even the very best scientists make them from time to time. But it's quite another to loudly declare that you've made one of the most momentous discoveries in the history of science before the glare of the world's media (and all before your paper has even been peer-reviewed). There are several lessons to draw from this incident, most obvious that humility is advised when presenting a result that could have truly profound implications.

But perhaps most important is that the fierce desire to discover, the desire to be first, can be dangerous. When properly harnessed, it drives science forward, but when allowed to run rampant, it can lead to wishful thinking, the suppression of awkward evidence, and, in BICPE2's case, precarious shortcuts. As Richard Feynman warned, "The first principle is that you must not fool yourself—and you are the easiest person to fool." These are words that should be framed above every scientist's lab bench.

When it comes to the anomalies we've been discussing, Feynman's warning is all the more crucial. As we've seen, anomalies can be heralds of revolutionary breakthroughs in our understanding of nature, but they can also lead science astray, taking entire fields up blind alleys and ruining careers. Broadly speaking, there are three ways an anomaly can go bad. The most frightening, and often most difficult to notice, are the sorts of straight-up experimental errors that led to the BICEP2 debacle. In other rarer cases, theorists can get their calculations wrong, giving the false impression that a measurement is in conflict with the prevailing theory.

We'll get to those soon, but before we do, let's consider perhaps the most common cause of anomalies, which are also fortunately the easiest to avoid.

Lies, Damn Lies, and Statistics

In December 2015, just as the Large Hadron Collider powered down for its scheduled winter break, the ATLAS and CMS collaborations, the international teams that operate two of the collider's giant detectors, presented an intriguing pair of results. After scouring trillions upon trillions of high-energy proton smashups, they had each noticed a tantalizing bump in a graph.

Now, there is nothing that gets particle physicists more hot and bothered than a bump in a graph. Such deviations often presage the discovery of a new particle. In this case, the putative new particle was a whopper, weighing in at 750 GeV, or around six times the mass of the Higgs boson, a particle whose recent discovery had made headlines around the world. At the time, a drought of discoveries since the Higgs had begun to make many fear that the Large Hadron Collider—constructed at enormous cost and bearing outsized expectations—might yield only a subatomic desert. And so the first hints of a new particle were extremely welcome indeed.

Within just a few weeks of the presentation, more than five hundred papers had been penned by theorists attempting to explain the bump in terms of this or that theory. A common proposal suggested it was the first member of a whole zoo of super-particles predicted by the grand theoretical paradigm known as supersymmetry. If that were true, then we were at the start of the biggest breakthrough in fundamental physics since the 1980s (or March of the previous year, if you count the discredited BICEP2 result).

In any case, lots of physicists were very excited indeed, and the excitement reached fever pitch at the International Conference on High Energy Physics, held in Chicago in August 2016. Both the ATLAS and the CMS teams were scheduled to present new studies of the bump, adding additional data recorded that year, promising to turn their tentative hint into a full-blown discovery.

But to the dismay of many in the community, when they revealed their crucial graph, the bump was nowhere to be seen. With more data it had simply melted away. The oasis in the desert had proved to be nothing more than a mirage.

So, what had gone wrong? Had the experimenters ballsed things up? Was there something wrong with the detectors?

Well, actually, no, nothing had gone wrong. When ATLAS and CMS presented their results in December, they had been careful to state that the statistical significance of their result was far below the threshold required to declare a true discovery. Unlike the BICEP2 team, they had made no grand claims. They had merely presented the results as they were. If anyone was to blame, it was theoretical physicists, the overexcitable little dears, who were so desperate for signs of something new that they were prepared to jump on any bump, no matter how insubstantial.

The bump, it turned out, had been nothing more than a cruel statistical fluke—a wobble in the data of the kind that happens from time to time, purely by chance. Understanding the whole issue of statistical significance is critical to our story of anomalies, and this peculiar episode is a great way to get our heads around how and why statistics can occasionally fool us if we don't have our guard up.

The first thing to understand is that all measurements come with "uncertainties" or "errors," two words often used interchangeably. The uncertainty on a measurement is an expression of the precision with which we think we have measured a particular quantity. Uncertainties come in two key types. There are *statistical uncertainties* and *systematic uncertainties*.

Let's start with statistical uncertainty. To draw on the classic example, imagine I gave you a coin and asked you to determine if that coin is fair, or to put it more formally, if the probability of that coin coming up heads or tails is equal. To test this, you toss the coin twice and get one head and one tail. On the face of it, this might suggest that the coin is fair. But you probably have the feeling that we can't really be sure from only two tosses. Indeed, like all measurements, this one comes with an uncertainty, and with only two coin tosses this uncertainty is large. I'll spare you the math and tell you that it's around 26 percent. We might therefore express our measurement of the probability of getting a head in this situation as

$$p(\text{heads}) = 50\% \pm 26\%$$

What does this mean? Well, according to the standard statistical definition of uncertainty, this means that there is around a 68 percent chance that the *true* probability of the coin coming up heads lies within one uncertainty of our measured value. Or put another way, we can say with 68 percent confidence that the probability of getting a head is between 24 percent and 76 percent. Our two throws haven't given us enough data to say whether our coin is fair or not. We need more evidence.

So now let's imagine you toss the coin a hundred times and get fifty-two heads and forty-eight tails. In this case, your measurement would be expressed as

$$p(\text{heads}) = 52\% \pm 6\%$$

Because you have added more data, the size of the uncertainty has shrunk. Now you can say that there is a 68 percent chance that the probability of our coin coming up heads lies between 46 percent and 58 percent. This is consistent with it being a fair coin, although there's still a reasonably large range. This is a key characteristic of statistical uncertainties: they get smaller with more data. Now let's imagine you really went to town and tossed the coin 10,000 times, getting 5,301 heads and 4,699 tails. Now things get interesting:

$$p(\text{heads}) = 53.0\% \pm 0.9\%$$

Your measured value is now more than three times the uncertainty away from 50 percent. In statistics, uncertainty is referred to by the Greek letter sigma, and here we have what's known as a three-sigma result, hinting that the coin may in fact be biased. It means that statistically, there is less than a one-in-a-thousand chance that you would get a result this far from evens—5,301 heads, 4,699 tails—from a fair coin.

As it happens, this was roughly the statistical precision of the bumps seen at the LHC in 2015. In both cases, the wobbles in the two graphs were just over three uncertainties away from the number of particles you would expect from random background processes. This meant that, in practical terms, there was only a one-in-a-thousand chance of the data randomly wobbling up in a way that would give the illusion of a new particle.

On the face of it, three-sigma evidence might sound pretty convincing; a 999-in-1,000 chance of having discovered something new is pretty good odds, right?

Okay, but now imagine that I told you that you were only one of a thousand people I had given a coin to, and that they had all done similar experiments, and you were the only one who had gotten a result that crossed the three-sigma threshold. Knowing that you were only one of a large number of measurements might make you wonder whether you just got your result by dumb luck. In other words, the coin really was fair, but just by random chance it came up with more heads than you might naively expect.

This, too, is more or less what happened at the LHC in 2015. Collider experiments don't just search for one type of particle in one way; they make literally thousands of measurements and searches. This means that over time you should *expect* a few of these to give you three-sigma fluctuations, just by chance. This is why no good particle physicist trusts a three-sigma result. The gold standard to declare a discovery is five uncertainties, or "five sigma," from your theoretical expectation. At that level, there is less than a 1-in-1.7-million chance of the result being a statistical fluke. And while we make thousands of measurements at the LHC, we certainly don't make millions.

Let's revisit the coin example one more time. Now that there's three-sigma evidence that the coin is biased, a load of theorists have gotten very excited and you have been given a large grant to support a hugely ambitious experiment to toss the coin more than 10 million times. Of course, now you're a tenured professor and you don't have time for such grunt work, so you hand the task to your underpaid team of PhD students and postdocs. After more than a year of round-the-clock tossing, you are finally ready to present your team's result to the world: 5,003,421 heads and 4,996,579 tails, giving

$$p(\text{heads}) = 50.03\% \pm 0.04\%$$

Because you have added more data, the uncertainty has shrunk and your three-sigma anomaly has vanished. The coin really is fair, to within a precision of just 0.04 percent. The original three-sigma result was just a fluke, nothing more. Alas, no Nobel Prize for you. You have been fooled by statistics.

This kind of thing happens in science all the time. It is especially common when you are dealing with limited data sets. The solution, in every case, is to take as much data as you can *and* try not to get overexcited when you see a three-sigma result. In particle physics we have a rather arbitrary standard by which we judge results like this: three sigma is regarded as "evidence"—that is the point at which it's worth paying attention to an effect, but long before you think about booking a ticket to Stockholm. Meanwhile, five sigma is an "observation" or "discovery." This five-sigma gold standard is there to protect us against being fooled by statistics.

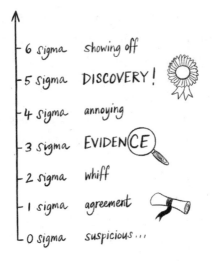

Here's a trusty scale I've invented to help think about an anomaly and its level of disagreement with prevailing theory:

As I said, these sorts of statistical flukes are straightforward to deal with; they always get found out in the end. However, there are other types of errors that can be much harder to untangle.

Dark Seasons

In July 2014, I traveled to sunny Valencia, Spain, to attend the biggest particle physics event of the year, the International Conference on High Energy Physics. After growing somewhat bored listening to report after report on the failure (so far) to find any signs of new particles at the Large Hadron Collider, I decided to wander into the astroparticle physics* session to see

---

* Astroparticle physics is the study of particles that arrive at Earth from outer space.

what might be going on. I was rewarded with one of the most ferocious and entertaining scientific arguments I've ever witnessed, the physics equivalent of a no-holds-barred wrestling match, only without any actual physical contact.

This rather forceful exchange had been triggered by a presentation from the team behind an experiment buried inside the Gran Sasso mountain in central Italy, who for several years had been making a highly controversial claim: that they had discovered the first direct evidence for dark matter.

As we've seen, dark matter dominates the physical universe. Five times more abundant than the ordinary matter that makes up galaxies, stars, and you and me, its gravity binds the universe together and was crucial in the formation of structure after the big bang. However, despite a wealth of astronomical evidence that it exists, we have no idea what it's made from. Detecting dark matter particles directly is therefore one of the great missions of modern physics.

So, any experiment claiming to have spied dark matter particles was inevitably going to attract a lot of scrutiny. Known as DAMA/LIBRA, the experiment in question had been designed to search for these particles using crystals of sodium and iodine. The idea was that as the Earth plows through the invisible cloud of dark matter that engulfs our galaxy, occasionally a dark matter particle would bump into an atom in one of the crystals, producing a flicker of light. Since 1998, the team had been reporting just such signals, and what's more they had found overwhelming evidence of what is regarded as the smoking gun of dark matter detection: a signal that rises and falls with the seasons.

Why should the seasons have anything to do with dark matter? Well, the Sun is currently orbiting the Milky Way at 828,000 kilometers per hour, and while the galaxy is rotating, its associated dark matter cloud is stationary. This should create an 828,000-kilometer-per-hour dark matter wind blowing continuously through the solar system. Meanwhile, the Earth is orbiting the Sun at 107,000 kilometers per hour. On around June 2 every year the Earth's motion around the Sun is most aligned with the Sun's motion around the galaxy, creating a particularly strong dark matter headwind. Six months later, in early December, the Earth is moving in the opposite direction to the Sun's orbit, and so experiences a slightly gentler dark matter breeze. Just as your front gets wetter quicker when running into the wind during a rainstorm, you'd expect to get hit by more dark matter when you're moving into the wind than when you're going in the opposite direction. So, if you see a rate of dark matter interactions that rises and falls with the seasons, that's a clear indication that you're seeing genuine dark matter particles.

Not only had the team behind DAMA/LIBRA been reporting a seasonal signal since 1998, but their signal also peaked in June, the exact time that the Earth is moving fastest with respect to the unseen dark matter cloud. On the face of it, it looked like a slam-dunk discovery. However, a number of subsequent experiments that should have been sensitive to the same dark matter particles had since come up empty-handed.

After the presentation concluded, a member of the audience got up to ask why DAMA/LIBRA refused to make their data public so others could cross-check their results. The speaker responded that it was their proprietary data and in any case nonexperts wouldn't be able to make sense of it without a detailed understanding of the experiment. This left the questioner, and many others in the room, rather unsatisfied to say the least, leading to a heated exchange that was eventually brought to an end by the session chair.

A decade later, very few scientists outside the DAMA/LIBRA team take their result seriously. A range of far more sensitive experiments have all but ruled out their claimed signal, including two experiments that essentially replicated their setup in other laboratories. However, the DAMA/LIBRA team remains defiant: as far as they are concerned, their result stands; it's the other experiments that have messed up their measurements.

So, what could have gone wrong with DAMA/LIBRA? Well, on the face of it, it looks like a classic example of the second key type of uncertainty—a *systematic error*. A systematic error is an effect that can bias the result of an experiment, or even completely spoof a signal. Dealing with such systematic errors is arguably the most important and difficult job of an experimental physicist; it is only by questioning every part of your experiment and analysis procedure that you build confidence in the final result—a statement that is all the truer when claiming a major discovery. However, finding systematic errors can be an extremely difficult task. Even the best scientists can miss effects simply because they never thought of them, or because of a flaw in how an effect is estimated.

One notorious case is that of the OPERA, a particle detector housed in the same subterranean lab inside the Gran Sasso mountain. OPERA studied neutrinos, particles famed for their ability to pass through even the densest material. For the purposes of their study, a beam of neutrinos was created by a particle accelerator at CERN, near Geneva, then fired *under* the Alps and down the length of Italy to Gran Sasso, where it was measured by OPERA's detector. In 2011, they reported that these neutrinos appear to have traveled through the Earth faster than the speed of light.

The speed of light is an unbreakable speed limit according to Einstein's theory of relativity, so the consequences of such a discovery, if true, would be profound. To be fair to the OPERA team, when they presented their results, they directly invited other scientists to point out the mistakes in their measurement. They made no grand claims that they had proven Einstein wrong.

Nevertheless, when the issue was eventually traced to a cable that hadn't been plugged in properly, the team operating the experiment faced ridicule in some quarters, rather unfairly, in my view. These experiments are incredibly complex, and it's always possible to miss something despite your best efforts, particularly when you have a shed load of cables to check.

But what about DAMA/LIBRA? How to explain the fact that they had allegedly seen a clear annual signal, year after year after year? A potential answer came from an independent team of Italian physicists in 2020, who noticed that DAMA/LIBRA was doing something rather dangerous when analyzing their data.

One of the biggest challenges of dark matter experiments is beating down background noise from natural radioactivity, which can easily swamp a faint dark matter signal, like a loud vacuum cleaner making listening to the telly nigh on impossible. Experimenters therefore go to extreme lengths to eliminate all sources of radioactive contamination. But even after creating as radiation-free an environment as possible, there will always be some residual background that you can't get rid of. The trick then is to accurately estimate this leftover radioactivity and take it into account when you make your measurement. The way DAMA/LIBRA dealt with this was to take an average of the background radiation rate over the course of a year and then subtract it from their measured dark matter signal every September.

However, doing something to your data once per year is a

terrible idea when you are trying to measure a signal with an annual cycle. In essence, the DAMA/LIBRA team risked injecting an artificial yearly wobble into their data, which under the right circumstances could make you think you'd discovered dark matter.

Nicola Rossi, a physicist independent of DAMA/LIBRA, was part of a four-person team who first suggested this risky procedure might be the cause of the claimed dark matter discovery. Their aim, he later told me, was to encourage DAMA/LIBRA to reveal a key bit of information that would allow the outside world to check whether this effect could be at the root of the mystery. So far at least, the experimental team hasn't taken them up on their challenge.

"They are good physicists, individually," he said. In the early days, the DAMA/LIBRA team members were pioneers of dark matter detection, and their early results were generally believed. "The problem is when they come together, you get groupthink," Rossi offered. Now they struggle to get published in mainstream journals.

More recently, an experiment in Korea called COSINE-100 that uses the same detection technology applied DAMA/LIBRA's method for dealing with background radiation, and they found that it did indeed induce an artificial annual signal—the kind the Italian team claimed was evidence of dark matter. So, from the outside at least, it looks as though systematic error were to blame for DAMA/LIBRA's finding. However, until DAMA/LIBRA is prepared to reveal the key bits of data, we can only speculate. (I approached members of the DAMA/LIBRA team to talk about the anomaly but was met with a wall of silence.)

If there's a lesson in this, it is about humility. The OPERA team checked and checked their result for an explanation, and when they couldn't find one, they put it out into the world and invited people to help them find their mistake. Such openness to criticism is essential if you really want to discover something new about nature. Hoarding your data and sticking to your guns in the face of criticism—that, on the other hand, is a recipe for distrust and doubt.

Mistakes in science are inevitable. The important thing is to remain humble enough to admit when you've made one.

With these cautionary words still ringing in our ears, it's time to dive headlong into the anomalies of today—the findings threatening to reshape our understandings of the cosmos. Where better to start than at the bottom of the Earth.

Chapter 4

Rising from the Ice

*Could impossible particles emerging from beneath the
Antarctic ice be clues to a new subatomic world?*

L inda Cremonesi's first day at work did not begin well. As
a newly qualified postdoctoral researcher who could now
proudly place the letters "Dr." in front of her name, and
keen to make a good impression, she had arrived at University
College London (UCL) bright and early at 9:00 a.m., only to
find the Department of Physics deserted. As she would soon
learn, most of her colleagues preferred a rather more leisurely
start to the working week. It also didn't help that her new boss,
Professor Ryan Nichol, was on an expedition in Greenland and
had forgotten to tell anyone she was coming.

Feeling slightly lost, she wandered the empty corridors trying to figure out where she was supposed to be, until the passing head of the department, Mark Lancaster, took pity on her
and guided her to a desk.

Cremonesi had just completed her PhD working on an
experiment in Japan known as T2K, which studies neutrinos.
Neutrinos have long been a subject of fascination for physicists, partly thanks to their spectral ability to pass through
thousands of kilometers of solid rock as if it were naught but
air, but also for their tendency to seamlessly morph from one
type into another as they travel.

The neutrinos that Cremonesi had studied in Japan were produced by a particle accelerator on the Pacific coast and fired through 295 kilometers of earth and rock to reach a giant subterranean detector on the other side of the country. This detector was essentially a huge tank containing fifty thousand tons of ultrapure water, lined with glistening golden orbs whose job was to detect the faint flickers of light produced when a neutrino bumped into a water molecule.

The experiment that she had come to UCL to work on was a completely different beast.

Cremonesi had been at a meeting with her collaborators in Japan when she'd heard about the job on offer at UCL. The successful candidate would be working on ANITA, a neutrino detector, like T2K, but instead of detecting light using small golden orbs, it searched for short blasts of radio waves using a huge antenna suspended from a high-altitude balloon. And instead of using a water tank as a target for the neutrinos, ANITA used the entire continent of Antarctica. When she read that the job would include travel to the frozen southern continent, she knew she had to apply.

Fast-forward a few months, and here she was at UCL, at the start of a scientific journey that would take her to one of the most extreme environments on Earth—and to the frontier of particle physics.

At first, Cremonesi felt a little out of her depth. While she came from a particle physics background, most of her fifty-odd collaborators on ANITA were astrophysicists. Moreover, the experiment was unlike anything she'd worked on before. She would have to learn a completely new way of detecting neutrinos, more or less from scratch.

ANITA, which stands for Antarctic Impulsive Transient

Antenna, is a radio antenna about the size of a double-decker bus, designed to search for ultra-high-energy neutrinos that rain down on Earth from outer space. The instrument, whose forty-eight gleaming white radio horns make it look rather like a giant sound system, floats high in the stratosphere, suspended from a NASA-operated balloon, slowly circling the continent on the circumpolar winds. If an ultra-high-energy neutrino hits the ice sheet of Antarctica, forty kilometers below, it will emit a pulse of radio waves that ANITA should be able to detect. Seeing evidence for such ultra-high-energy neutrinos would be a huge coup, potentially opening a window to some of the most violent objects in the cosmos.

To understand why the ANITA team goes to such extremes to spy these neutrinos, we first need to step back and think a little more about these strange particles. Neutrinos are often described as "elusive" or "ghostly" due to their ability to fly through solid objects. They are created in such vast quantities by the nuclear reactions that power the Sun that in the time it takes you to read this sentence, around ten trillion will have passed straight through you. You are blissfully unaware of this relentless subatomic onslaught thanks in large part to the fact that neutrinos have no electric charge. No electric charge means that they can't push or pull on other electrically charged particles (like the electrons that make up the atoms in your body). Nor do neutrinos interact via the strong force—the force that holds the atomic nucleus together—so they don't get deflected by nuclei much either. In fact, the *only* way that a neutrino can exert any influence on ordinary matter is via the so-called weak force.

As the name suggests, the weak force is pifflingly feeble compared with the electromagnetic and strong forces. This means that the chances of a neutrino knocking into an ordinary atom are extremely low. This is why they can pass through rock, mountain ranges, even entire planets, without exchanging so much as a flirtatious wink with an atom. Incredibly, it would take a block of lead a light-year thick to have even a fifty-fifty chance of stopping a neutrino.

While making them fiendishly hard to detect, neutrinos' indifference to matter also makes them a potentially unique source of information about the universe. Take, for example, the burning core of our Sun. A photon produced in the solar core will endlessly pinball off the electrically charged electrons and nuclei that make up the churning, superheated bulk of the Sun, taking tens of thousands of years to break out from the surface and escape into space as sunlight, by which point any information it might have once carried about the conditions in the solar core has been lost. A neutrino, on the other hand, will zip unimpeded through the Sun like light through glass. If it can be detected on Earth, it can give scientists a window directly into the heart of our closest star.*

The Sun is a terrifyingly powerful object compared with us humans, but it pales into utter insignificance next to the astrophysical monsters that ANITA is attempting to study. These awesome objects include so-called active galactic nuclei—supermassive black holes billions of times heavier than the Sun that can blast cataclysmic beams of energy out into space, powerful enough to dismember entire galaxies. Such objects

---

* An experiment in Italy known as Borexino did just that, recently reporting the first direct evidence for the nuclear reactions that power the stars. You can read more about this remarkable experiment in my book *How to Make an Apple Pie from Scratch.*

act as natural particle accelerators, that make the Large Hadron Collider look like a toy train set.

We'll get back to ANITA in a moment, but first it's helpful to get a feel for the various energy scales at play. The basic unit of energy used by most particle physicists and astrophysicists is the electron volt, usually written as eV. One eV is the amount of energy an electron would get if you accelerated it using a one-volt battery. It's a useful enough definition, but an electron volt is tiddlingly insignificant in particle physics terms, where we generally deal with millions, billions, or trillions of electron volts, written as MeV, GeV, and TeV. For instance, if you were to take an electron at rest and convert its mass into pure energy, you'd get half a million electron volts out, or 0.5 MeV. Meanwhile, the LHC, the world's most powerful particle collider, can accelerate protons to a whopping seven trillion electron volts, or 7 TeV. Even that is peanuts next to the energies of the neutrinos that ANITA is searching for.

Earth is under constant bombardment by particles raining down from outer space, collectively known as cosmic rays. Most cosmic rays are protons or atomic nuclei, with average energies of a few hundred MeV. However, every so often, a cosmic ray carrying a gigantic amount of energy smacks into Earth. These ultra-high-energy cosmic rays carry more than a million trillion electron volts—an exaelectron volt, or EeV for short. The most potent cosmic ray ever detected crashed into Earth on October 15, 1991. With a truly stupendous energy of 300 EeV, it became known as the Oh-My-God particle. To give you a sense of what this dizzyingly huge number represents, if a baseball moving at the same speed as the OMG particle crashed into Earth, it would release over a thousand times more energy than the asteroid that wiped out the dinosaurs.

The Oh-My-God particle presented physicists with a conundrum. While it was possible to imagine how a particle might get whipped up to such enormous energies by a supermassive black hole, what was puzzling was that it had survived the journey to Earth at all. Intergalactic space is filled with a freezing haze of microwave photons—the faded remnant of the light left over from the fireball of the big bang. Every cubic centimeter of space contains just over four hundred of these photons, each carrying a vanishingly tiny amount of energy—a mere thousandth of an electron volt. Compared with the Oh-My-God particle, these photons are like a buzzing cloud of gnats in the path of a speeding juggernaut.

But while these photons might be triflingly low in energy for an observer at rest, from the point of view of the Oh-My-God particle they represent a lethal hail of subatomic bullets. I'm sure at some point you'll have been sitting in the front of a car speeding down a freeway when an unfortunate fly smacks into the windshield with a surprisingly loud splat. Well, imagine that instead of driving at seventy miles per hour, you were going at 99 percent of the speed of light. That little fly wouldn't just make a splatter on your windshield; the impact would vaporize your car.

This effect means that protons with energies larger than about 50 EeV shouldn't be able to reach us from galaxies outside our local neighborhood, which is where most supermassive black holes can be found. But evidently, in 1991, one did. The question is how?

One explanation for the Oh-My-God particle is that, rather than a proton, it was an iron nucleus, whose larger weight allowed it to plow through the photon hail. However, iron nuclei make up only a tiny fraction of the ultra-high-energy cosmic rays that are thought to be produced in the distant reaches of the cosmos. Also, protons and nuclei get bent by magnetic fields that thread the cosmos, which means they end up traveling along paths that weave and wind their way around galaxies, making it impossible to tell where they originally came from once they're detected on Earth.

All this means that if we want to understand the extreme objects that create ultra-high-energy cosmic rays, we need to seek out different messengers.

This is where neutrinos come in. Since a neutrino has no electric charge, it doesn't interact with photons, and so it can breeze through the microwave haze without any bother, no flies on its proverbial windshield. Similarly, their lack of electric charge means they aren't affected by magnetic fields, and so they zip through the cosmos in straight lines, making it possible to figure out where in the sky they were produced. And since astrophysicists expect supermassive black holes to produce ultra-high-energy neutrinos, along with other charged particles, they represent perhaps the best way to get a handle on what's happening inside these terrifying, extremely distant objects.

The only trouble is that these ultra-high-energy neutrinos are expected to be very rare. Meaning you need a really, *really* big detector to have any chance of ever seeing them. That's where Antarctica comes in. With ANITA, the entire continent becomes the detector.

Here's how it works. We've said that neutrinos can pass straight through entire planets with ease. This is actually only true for low-energy neutrinos. When you get up to the incredible energies of the neutrinos that ANITA is searching for, the weak force stops being weak and (for reasons we can skip past here) becomes just as strong as the electromagnetic force.* As a result, these neutrinos can't penetrate the whole Earth, and instead have a comparatively high chance of smacking into an atom in the Antarctic ice. When this happens, the huge energy of the neutrino gets converted into a shower of electrically charged particles that fly through the ice and create a blast of radio waves. When these radio waves then escape from the ice, they can be detected by ANITA as it floats high above the continent on a giant helium balloon.

Or at least that was the idea. Making it work in reality was a whole other matter, as Cremonesi would soon discover.

To the End of the Earth

Linda Cremonesi's love affair with physics began when she left her childhood home in a small village in northern Italy to study at the University of Milan. While at university, she spent a year overseas as an Erasmus exchange student at Queen Mary University in London. It would prove to be a turning point, both for her future career and for her personal life. Having experi-

---

* If you're interested in why the weak force gets strong at high energies, it's ultimately due to the particles that transmit the weak force—the so-called W and Z bosons. The W and Z have huge masses, 80 and 90 GeV, respectively, meaning that they take a lot of energy to produce directly. This is ultimately what makes the weak force weak: since the W and Z are so heavy, low-energy particles don't have enough energy to produce them. This is different from the electromagnetic force, whose particles—photons—have no mass and so are easy to produce regardless of how much energy you have. However, ultra-high-energy neutrinos have so much oomph that they *can* create W and Z bosons when they smash into ordinary matter, turning the weak force into a strong one.

enced a lot of homophobia in Italy, she found London to be a breath of fresh air, somewhere she could finally be herself. "So, I was like, you know what? I'm just gonna stay."

After moving to the U.K. in 2009, she bounced back and forth between Queen Mary in London's East End and UCL, five miles to the west in the rather more genteel Bloomsbury district. First came a master's degree at UCL (supervised by Ryan Nichol), then a PhD at Queen Mary, and then back to UCL again for her first postdoc.

Now on ANITA, she would be part of an international team of around fifty people, who were rather inconveniently spread over multiple time zones, from Taiwan to Hawaii. This entailed meetings that started at 9:00 p.m. in the U.K. and often ran late into the evening. To start off with, Cremonesi's role would be running simulations in preparation for the upcoming fourth and final flight of ANITA, which was scheduled to launch in December 2016. Soon, though, she would have the opportunity to get her hands dirty.

Before being shipped to Antarctica, ANITA had to undergo stress testing at NASA's balloon facility, just outside Palestine, Texas. Married on July 2, 2016, Cremonesi got only three days to spend with her new (and presumably very understanding) wife before flying out to the United States for more than a month.

It's hard to imagine a more contrasting environment to the frozen southern continent where ANITA would soon reside than rural Texas. As Nichol put it, the small town is "ridiculously hot" in the summer, with average highs of thirty-five degrees Celsius (ninety-five degrees Fahrenheit), and is surrounded by alligator-infested swamps. By the time Cremonesi arrived, most of the team had already been there for a few weeks.

Their first job was to assemble ANITA, which had been shipped to Palestine in boxes from Hawaii. Next it was suspended from the ceiling of a huge hangar to simulate being attached to its balloon. Then the experiment's instrument box was put in a very cold, low-pressure chamber to see if it could survive the conditions high in the Antarctic stratosphere. These hurdles cleared, everything was finally packed back into boxes, ready to be shipped south.

While Cremonesi and her colleagues were laboring in the sweltering Texas heat, the ANITA collaboration published a new analysis of the data recorded during the experiment's maiden flight a decade earlier. The result, belated though it was, would seriously raise the stakes for their coming mission.

As we've seen, ANITA was designed to detect radio pulses produced by ultra-high-energy neutrinos as they collide with the Antarctic ice. Regrettably, analyzing the data from that first flight revealed no signs of the neutrinos they sought. But they did find something they hadn't anticipated: radio pulses that appeared to come from ultra-high-energy cosmic rays— charged particles like protons and nuclei—as they collided with Earth's atmosphere.

On reflection, this wasn't altogether surprising: ultra-high-energy cosmic rays hit the atmosphere fairly frequently. But it wasn't just the appearance of these cosmic arrivals that caught their attention. On studying the pulses in more detail, they found evidence of sixteen separate cosmic ray events. Fifteen of them behaved as expected, passing through Earth's atmosphere before being detected. One of them, however, appeared to have traveled *upward,* from below ANITA, as if it had emerged from the ice sheet itself.

According to the known laws of physics, such a thing should not be possible. Cosmic rays can penetrate only a short distance through the Earth before being brought to a halt by collisions with ordinary atoms. The direction of this upward-going cosmic ray implied it had traveled thousands of kilometers *through* the Earth, which, given its enormous energy, should be impossible, even for a neutrino.

If the signal couldn't be explained using known physics, what could have caused it? There were many tantalizing possibilities. Was it evidence of a so-called super-particle, of the kind predicted by the popular extension of the standard model known as supersymmetry? Or was it a messenger from the invisible dark universe, a particle that might be linked to some of physics' deepest mysteries?

One swallow doesn't make a summer, and while an intriguing finding, a lone cosmic ray was a long way from proof that the laws of physics were breaking down. The question for Cremonesi and her collaborators now was, would ANITA's fourth flight reveal any more of these strange upward-going particles?

After two months in Texas, Cremonesi flew back to the U.K., where she immediately had to start getting ready for the trip to Antarctica. Before being allowed to travel, she would have to pass a grueling series of medical tests to check for any lurking health conditions that could potentially flare up during her trip. In Antarctica there would be only two doctors for the whole research base of more than a thousand people. The nearest hospital would be at least a five-hour flight away in New Zealand, with only one flight leaving a week, weather permitting. Going to Antarctica is a bit like going into space, and once upon a time you had to have your wisdom teeth removed before traveling. Fortunately, by the end of October, Cremonesi came through the tests with the all clear. "It was good to know I didn't have any horrible diseases," she said.

She would find out when she'd be leaving with only two or

three days' notice. The call came just a few days later at the start of November. Before she knew it, she was at Heathrow Airport checking in for the long flight to New Zealand.

Cremonesi spent two days in Christchurch on New Zealand's South Island preparing for the harsh conditions she would soon face. There she joined a group for a day of training, which mostly involved watching videos about life on the frozen continent. The need to conserve water meant you were urged to restrict yourself to two showers a week, with laundry once a week. "Everyone ended up a little bit smelly," Cremonesi said. Preservation of the pristine Antarctic environment was another focus. Littering was an absolute no-no. Even spilling tea on the ice required you to scoop it up; for anything bigger the research base had a spillage team who would get called in.

On day two in New Zealand, Cremonesi and her group received their extreme cold weather clothing: two highly insulated Canada goose jackets known as big red and little red (if red's not your color, tough luck), trousers, boots, gloves, goggles, and a hat. She was also handed a name tag to Velcro onto her jacket; everyone looks the same in their cold weather gear so the tag helps you recognize each other, but also, rather morbidly, it helps identify your body in the event of a fatal accident.

On the day of their departure, the group had to don their full cold weather gear despite it being the start of summer in New Zealand. As Cremonesi sat sweltering on the bus to the airport, she felt a rising mixture of excitement and nervousness. Her traveling companions were mostly scientists, but she was the only person from ANITA. It was a relatively small group of thirty or so, because, as it turned out, they would be flying with an unexpected guest. Their ride was a hulking four-engine U.S. Air Force military transport plane, and their seats were fitted around a helicopter, which took up most of the main compartment.

Cremonesi found herself sitting with a group of biologists on their way to study fish off the Antarctic coast, but the deafening roar in the spartan cabin made conversation difficult. Most people sat reading, listening to music, or tucking into their prepacked meals. The crew handed out earplugs to help ease the flight, which, despite the noise and the cold, was at least fairly smooth.

Five hours later, the plane touched down on the compacted snow runway of Williams Field, an airstrip atop an eighty-meter-thick sheet of permanent sea ice floating on the Southern Ocean. When Cremonesi stepped off the plane, the view took her breath away. A dazzlingly white flat landscape stretched out in every direction. To the southeast was the continent of Antarctica, the ramparts of the Transantarctic Mountains breaking through the thick ice, its white peaks shining bright in the perpetual summer sun. To the north lay Ross Island, dominated by the cone of Mount Erebus, the southernmost volcano on Earth. Their final destination was about five miles due west— a promontory of dark rock thrust south from the main body of the island, at the end of which was McMurdo Station, Cremonesi's home for the next month.

They were driven to McMurdo on Ivan the Terra Bus, a chunky red shuttle with tires as tall as a person. Ivan was a particular favorite of the residents of McMurdo thanks to its ability to speed along the compacted snow roads. "If you got to ride on Ivan, it was a good day," Cremonesi said. McMurdo is a U.S. research station and the largest settlement on Antarctica, serving as a logistics base for half the continent.

On arrival, Cremonesi found herself in a small town of

utilitarian metal buildings connected by dirt roads, nestled between a hill of dark volcanic rock and the permanent sea ice. On disembarking from Ivan, the group were sent straight for a training session to learn about their new home. Life at McMurdo was highly regimented. Cremonesi compared it to being in the military. Breakfast, lunch, and dinner were served at fixed times every day, you worked weekends, with one Sunday off every two weeks, everyone shared a room (Cremonesi was happy to find she was sharing with a colleague from ANITA), and everyone had a job to do.

The population of McMurdo reaches around twelve hundred in the summer, two hundred of whom are scientists, with the rest there to keep the scientists alive and reasonably comfortable. There are cooks, cleaners, doctors, nurses, firefighters, shuttle bus drivers, and engineers who look after the power and heating. The place is a strange blend between gritty mining community and U.S. college town, populated by a mix of gruff old miner types, who spend their time digging holes in the ice and driving big trucks, and hippieish U.S. college kids on gap years, who get six months in Antarctica of tax-free pay in exchange for doing the more menial jobs like cleaning and running the canteen.

As a twentysomething herself at the time, Cremonesi became good friends with some of the students working the canteen and bars. While you don't have much time off for socializing at McMurdo, there are three bars on the station—or what passes for a bar, actually a room where you can purchase a can of beer for the princely sum of three dollars.* When she wasn't work-

* The currency at McMurdo is the U.S. dollar as Cremonesi discovered to her dismay, having arrived with her wallet packed with New Zealand dollars.

ing on the experiment, she'd meet up with her new pals to play pool or shuffleboard, and every so often there would be a special event like a karaoke night. She recalled one rather surreal evening where she danced until 3:00 a.m., only to emerge into the dazzling summer sunlight for a short stagger back to her room to get some shut-eye before her 7:00 a.m. start. "I was young," she recalled. "I couldn't do that now."

At McMurdo, Cremonesi met up with her boss, Ryan Nichol. This was Nichol's fourth trip to Antarctica, having traveled out for each of ANITA's previous three flights. His first had been ten years earlier in 2006 as a young man fresh out of his PhD. Now happy to place himself in the "older scientist" category, Nichol's experience was rather different from Cremonesi's. He had two young girls back in London. On his previous trip in 2014, his wife had been expecting their second daughter, and she gave birth unexpectedly just a few days after Nichol caught the last flight out of Antarctica. Asked how his wife had felt about his being away while she was pregnant, he said, "She was fine about it, but if I hadn't got home in time, she would have been . . . less fine."

On Cremonesi's second day, bright and early at 7:30 a.m., she and Nichol boarded Ivan for a half-hour drive along the snow road to the NASA balloon station. The polar vortex that would carry ANITA around the southern continent had set up early that year, and the race was now on to get everything prepped for launch. The experiment was being assembled in one of the balloon station's two huge metal hangars, the tallest buildings in Antarctica, which (rather incredibly) rest on skis, allowing the seventy-ton buildings to be dragged to safer locations during winter so that they don't end up buried under snow. Once ANITA had been unpacked from its shipping crates, the team had to move quickly to assemble the structure, which gets bolted together a bit like a Meccano set, starting from the ground and working upward. At a certain point ANITA gets so big and heavy that it has to be suspended from a giant hook attached to the hangar's ceiling to allow work to continue. Next the forty-eight antennas and the computer control system are attached before the team finally switches the whole thing on.

It was all hands on deck during those intense days at the balloon station. Every minute counted: working hours were set by the bus timetable, which left McMurdo every day at 7:30 a.m. sharp (if you overslept, you weren't going to work), returning at 4:30 p.m. each afternoon. Once ANITA was assembled, there was a long list of preflight tests to get through, and to make matters even more challenging, not everything was working straight out of the box.

Life at the balloon station was rough. While the hangars were heated, there was little else in the way of creature comforts. When a snowstorm came in, there was rarely time to get back to McMurdo, leaving Cremonesi, Nichol, and the crew stranded at the balloon facility until the storm abated. With no bunkhouse, this meant sleeping in the hangar itself, with buckets strategically positioned in the corners to serve as makeshift toilets. Meals at least were provided, prepared in a half-buried cylindrical building that served as the station canteen. Cremonesi was delighted to find that they had been assigned an excellent cook who was able to work wonders with their rather meager supplies. "You never knew when you were going to get fresh vegetables," Cremonesi said. "We would all get really excited when a new flight came in because we would get 'freshies.'"

ANITA's launch window was set for late November. While the team was performing their final preflight tests, they got a salutary warning from the experiment ahead of them in the queue. The hangars are heated to about seventeen degrees Celsius, but outside it's a frigid minus ten, so cold that most computers struggle to boot up. To get around this obstacle, instruments must be switched on in the warmth of the hangar before being taken outside and attached to the balloon. The team ahead of them forgot this critical step, and by the time they realized their mistake, the instrument was already out on the ice, stubbornly refusing to boot up. Disastrously, after traveling to the ends of the Earth for a month of hard graft, one tiny mistake meant they missed their launch window.

Before ANITA's own launch, there would be one final hurdle to cross, the so-called hang test. This was the crucial moment when the launch vehicle—a gigantic truck with a crane on it—came and picked up their experiment. Nichol, Cremonesi, and the team then had to prove that their equipment could interface with the NASA balloon system. Happily, ANITA passed the hang test with flying colors. The moment had finally come: they were ready for liftoff.

Now they were at the mercy of the most important person at the balloon station: the weatherman. Like a rocket launch, NASA's long-duration balloons require perfect conditions: a steady breeze at ground level with the polar vortex circulating high above in the stratosphere. As they waited for the call, the team entered an anxious holding pattern. Their first chance came at the end of November. On launch days, a specially chartered bus drops you at the balloon station at 4:00 a.m., and so it was they arrived excited, if a little bleary-eyed, to see their experiment take flight. But by midday the weather had turned against them, and the launch had to be scrubbed at the last moment. A few days later, another opportunity came and went as conditions deteriorated partway through the day.

So followed yet another agonizing wait at McMurdo for the

weather to improve, until at last, on December 2, a third call came.

With stable weather conditions, Nichol, Cremonesi, and the team handed their precious payload over to the thirty-strong NASA balloon team. First, the five-meter-high ANITA instrument was carefully lifted out of the hangar and attached to the launch vehicle's huge crane by the NASA riggers—a "nerve jangling" moment, as Nichol put it. The launch vehicle then carried ANITA gingerly out onto the ice to meet the waiting balloon, which stood a few hundred meters off, a giant pearlescent teardrop rippling gently in the breeze.

The ANITA team could only watch from a distance as the NASA crew attached the long balloon cable to their experiment. The balloon itself, made of a gossamer-thin polyethylene film just two-hundredths of a millimeter thick, was still being inflated with the last of the 100 million liters of helium needed to loft it and ANITA into the stratosphere. The process took several hours. Using a handheld radio, Cremonesi and Nichol could listen in to the back-and-forth between the NASA crew.

Eventually, with everything in place the question "Ready to launch?" crackled over the radio. "Yes, ready to launch," came the reply.

"We had the full NASA experience," Cremonesi recalled with a smile, countdown and all: "Five, four, three, two, one. Release the balloon!" At that, the balloon jerked suddenly upward, rippling and rolling like a huge jellyfish as it rose into the air. As cheers and whooping broke out from across the balloon station, the launch vehicle to which ANITA was still tethered gradually adjusted its position as the cable straightened. At last, with the balloon high in the air and the cable taut, ANITA was set free to float off into the clear blue Antarctic sky.

Feeling a profound sense of relief and accomplishment, it should have been time for the team to a celebrate. But there was still work to be done. ANITA beams its data back to the ground using satellite relays, a painfully slow process, but the team would have a direct line of sight to the instrument over the next few hours, allowing for much faster data downloads than during its mission proper and providing a crucial opportunity to perform their final postlaunch tests. Nichol and Cremonesi were two of only four people who fully understood ANITA's data acquisition system and so had to remain at the balloon station communicating with ANITA until, at long last, it drifted out of sight.

Thankfully, the tests were passed. They were taking good data. The hunt for more impossible particles had begun.

Cremonesi had only a few short days left at McMurdo before a flight would carry her back to Christchurch, and then to London, half a world away. She was surprised by how sad she felt at the prospect of her imminent departure; she had grown used to her strange new life. "It was a little bit like Stockholm syndrome," she said. "I thought, 'This is my life now; this is my prison.'"

Her flight took off from Williams Field on December 7. She'd be home in time for Christmas. But ANITA's journey was just beginning. Somewhere, far off over the continent, the balloon floated high in the stratosphere, now swollen to monstrous proportions, a vast, translucent sphere the size of a football stadium, below which hung their comparatively tiny instrument, listening carefully for signals from the ice below.

The Impossible Particles

As Cremonesi and Nichol had been preparing to launch ANITA from the edge of Antarctica, the theoretical physicist Ian Shoemaker was nine thousand miles away in a seminar room at Fermilab. Just outside Chicago, Fermilab is the United States' national particle physics lab, a sprawling site that is home to numerous experiments, including the now defunct Tevatron, once the planet's most powerful collider. Then an assistant professor at the University of South Dakota, Shoemaker was visiting Fermilab to give a talk on neutrinos—a subject that had fascinated him throughout his career—outlining a theory he'd been working on involving an as-yet-hypothetical particle known as a sterile neutrino. His hope was that his proposal might explain some long-standing anomalies detected in neutrino experiments, including one at Fermilab.

As noted earlier, neutrinos interact with ordinary matter only via the weak force, which means they can pass through solid objects (including entire planets) with ease. A sterile neutrino, if such a thing exists at all, would be an even more elusive beastie. It doesn't even feel the feeble pull of the weak force, which makes it almost completely cut off from ordinary matter and thus impossible to detect directly.*

Now, you might very reasonably wonder why anyone would bother thinking about a particle that is impossible to detect. Wouldn't its existence become a matter of belief rather than

---

* The only way a sterile neutrino could interact with ordinary matter directly is via gravity, but gravity is so fantastically weak, even compared with the weak force, that it's all but irrelevant for particle physics experiments.

science, like ghosts or the tooth fairy? Well, even if you can't detect sterile neutrinos directly, there is a potential way to infer their existence. In Shoemaker's theory, it was possible for a sterile neutrino to *transform* into an ordinary neutrino as it traveled. If you spotted ordinary neutrinos simply popping up in a place they shouldn't be able to reach, that could suggest they had been produced by sterile neutrinos.

During the Q&A at the end of his talk, Shoemaker got hit by a curveball from a member of the audience, a prompt that left him stumped. "This is a cool scenario," the questioner said. "I wonder if it could explain this weird ANITA event?"

Weird ANITA event? Shoemaker had never heard of ANITA, let alone the mysterious event that the questioner was referring to. He could only respond with a shrug. But once he was back in South Dakota, the question lingered in his mind.

Digging into the scientific literature, Shoemaker and his postdoc John Cherry learned about the strange cosmic ray that ANITA had detected way back in 2006 and that seemed to have burst from beneath the Antarctic ice sheet, perplexing scientists.

The standard model of particle physics couldn't explain ANITA's upward-going cosmic ray, but, Shoemaker and Cherry realized, their sterile neutrino theory might. Unlike the ultra-high-energy neutrinos that ANITA sought to detect, a sterile neutrino would be able to pass straight through the Earth, even at enormous energies detected by ANITA. Then, as it came close to the surface of the ice sheet, it could convert into an ordinary neutrino, which would cause it to interact with the weak force and, in turn, convert it into a shower of charged particles and create the characteristic radio burst that ANITA had spotted.

It was an exciting possibility. If they were right, ANITA might have just found the first evidence of a particle that lies outside the standard model.

Shoemaker and Cherry penned a short paper explaining the idea and submitted it for publication in a scientific journal. While the journal was reviewing their work, their theory received an unexpected boost. The ANITA team announced that they had spotted a *second* upward-going cosmic ray, this time during the experiment's third flight in 2014.

With the announcement of a second impossible particle, other theorists started to get in on the act. Soon the literature was full of speculative solutions to the ANITA puzzle. Some linked the ANITA signals to dark matter, postulating that the events could be caused by heavy dark matter particles that had accumulated inside the Earth and were decaying or annihilating, producing (detectable) standard model particles. Others pointed to supersymmetry, a grand theoretical paradigm that proposed symmetry between particles like electrons, quarks, and neutrinos and force particles like photons and gluons. Supersymmetry had dominated theoretical physics for decades but had fallen out of favor in recent years, after a slew of searches at the Large Hadron Collider had come up empty-handed. Could ANITA resuscitate a theory that increasingly looked to be breathing its last?

In 2019, a paper by a group of theorists at Penn State University caused a blizzard of publicity that swept first down the hallways of university physics departments and then out into the wider media. They had carefully analyzed the ANITA events and found that there was a less than 1-in-3.5-million chance of them being caused by known standard model particles. What's more, they had uncovered supporting evidence of the anomaly in another Antarctic neutrino experiment: the mighty IceCube.

IceCube is a physics experiment on a vast scale: a cubic kilometer of ice at the South Pole threaded with thousands of light detectors, which hang like beads inside vertical boreholes that plunge a kilometer into the ice sheet. Like ANITA, IceCube searches for high-energy neutrinos arriving from outer space, which create bursts of light when they interact with ice. Poring over IceCube's data set, the Penn State team found three similar events. It suggested that ANITA's signal could be the real deal.

Things started to get silly, however, when a group of theorists published a paper suggesting that the ANITA signals could be evidence for a rather wild cosmological theory they had recently cooked up. In essence, they were proposing that along with our own universe, the big bang had spawned a mirror universe made of antimatter, one where time runs backward. A possible consequence of this theory, they conjectured, is that dark matter could be made of sterile neutrinos—which in turn could explain the ANITA results.

Even the scientists who wrote the paper would surely admit that this was a rather tenuous chain of conjectures. Unfortunately, for tabloid journalists looking to bash out a quick and dirty science story, the term "mirror universe" is like a red rag to a bull. This led to some ludicrously hyperbolic headlines, including "NASA Scientists Detect Parallel Universe 'Next to Ours' Where Time Runs Backwards" in the U.K.'s *Daily Star*,[*] along with an eerily similar headline in the *New York Post*.[†] It was enough to make you wonder whether both journalists

---

[*] Michael Moran, "NASA Scientists Detect Parallel Universe 'Next to Ours' Where Time Runs Backwards," *Daily Star*, May 17, 2020, www.dailystar.co.uk.

[†] Yaron Steinbuch, "NASA Scientists Detect Evidence of Parallel Universe Where Time Runs Backward," *New York Post*, May 19, 2020, nypost.com.

misheard the same story shouted by a drunk in a very noisy pub, possibly also in a parallel universe.

I must admit to being somewhat in awe at how gloriously wrong this headline was, pulling off the extraordinary feat of squeezing three glaring untruths into just twelve words, a hit rate that even Donald Trump would be proud of. (ANITA is not run by NASA scientists, they definitely didn't detect a parallel universe, and even if they had, it isn't next to ours but on the other side of the big bang, which is about as far away in both time and space as you could imagine.)

Be that as it may, the ANITA team was left somewhat bemused by the sudden flurry of interest in their work, particularly because the press storm came several years after they released their results. Ryan Nichol put it rather generously. "It was lockdown 2020 and everyone had gone a bit crazy," he said. He and his colleagues were careful not to feed the frenzy. They had simply stated that they had seen two signals they couldn't explain; it was left open as to whether this was the sign of something genuinely exciting or a quirk of their experiment.

Meanwhile, the theorist Ian Shoemaker continued to puzzle over the ANITA results, along with his onetime PhD adviser, Alex Kusenko. They'd started to wonder whether there might be a more mundane explanation for the anomaly. What if the two radio pulses that looked as if they came from cosmic rays emerging from beneath the ice had actually been produced by cosmic rays entering Earth's atmosphere from above? If the radio pulse had reflected off the ice, it could *look* as if it had come from below. Now in principle ANITA is able to tell direct and reflected radio pulses apart based on the shape of the pulse. Here's a rough sketch of the difference:

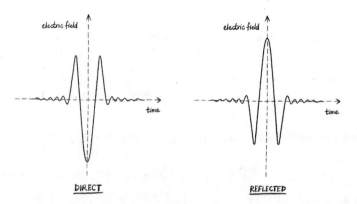

On the left is a radio pulse received directly by ANITA, while the pulse on the right is typical of one that's bounced off the ice. Notice their opposite orientations, a result of the well-established fact that reflection flips the "polarity" of a radio pulse.

Both the weird upward-going pulses seen by ANITA had an ordinary, un-reflected shape. So at first glance, it seemed impossible that they could be reflections. However, the Antarctic ice sheet isn't just a solid block of uniform ice. Could complicated reflections off subsurface features—perhaps hidden lakes of liquid water—have spoofed an upward-going signal?

It was an intriguing possibility, but when Shoemaker and Kusenko approached the ANITA team to discuss the idea, they got short shrift. After all, what right did two theoretical particle physicists have to pontificate about Antarctic ice formations from their cozy university offices? If they were to be taken seriously, they needed expert input. So they teamed up with some glaciologists to think about the issue in more detail.

It turned out there were a number of potential subsurface formations that could pull off the trick of reflecting an incoming radio pulse without flipping its polarity. Most promising were double layers of ice on top of hoar (ice crystals that form when water vapor in the air freezes on contact with the ground) or places where there was a change in density between buried layers of compacted snow known as firn. According to their calculations and previous radar surveys of Antarctica, such features were common across the continent, potentially explaining the ANITA events without the need to invoke exotic new physics. To find out whether they were right would require a dedicated survey of the mysterious subsurface features of Antarctica.

So far at least, no one has taken them up on their suggestion. Still, their paper was enough to provoke a response from the ANITA team, who conducted a detailed study of their own. Their conclusion? It didn't seem a likely explanation. They found no evidence of the necessary subsurface features at the locations of the two mysterious signals, and what's more, the strength of the signals they'd recorded appeared to be far too high to have come from reflections. Such a strong reflected signal would have to have come from particles with ten times more energy than first supposed. If such particles were really hitting Earth, they'd surely have been noticed elsewhere.

As all this was rumbling on, ANITA were sitting on a potential bombshell: the data recorded by Cremonesi, Nichol, and the team on ANITA's fourth and final flight. The data analysis neared completion, and the big question on everyone's minds was, would more upward-going particles be spotted?

The answer, when it came, was frustratingly inconclusive. Instead of drawing a blank, or detecting more particles emerging from the ice, they found four new events that appeared to come from just below the horizon, merely grazing the ice sheet. Since they had traveled only a short distance through the ice, no exotic explanation was required. Instead, they could be explained by the particles ANITA had been hunting for all along: ultra-high-energy neutrinos.

On a clear, bright spring morning, I sat down with Ryan Nichol in UCL's coffee room to puzzle over the rather confusing state of affairs. After four trips to Antarctica, years of data analysis, and dozens of theory papers attempting to explain ANITA's results, what did he think were the chances that they had really spotted the first signs of particles that lie outside our current understanding of nature?

"If I had to bet—I'm very cynical—I'd say the very vertical ones are down to some mundane bit of physics." Such skepticism is common, not to mention prudent, among experimental physicists: no one wants to get egg on their face by claiming to have broken the standard model, only for it to turn out to be an analytical cock-up. A key piece of evidence against an exciting explanation for the ANITA events is the fact that the IceCube experiment has failed to further corroborate their results.

For all his skepticism, Nichol freely admits that no one has yet satisfactorily explained ANITA's mysterious particles. And there are plenty of ways that exotic new physics could evade IceCube but show up in ANITA.

Theorists tend to be more bullish. As Shoemaker told me, "I often feel that experimental collaborations are almost embarrassed to have an anomaly. . . . But for theorists it's completely the opposite—it's the most exciting thing!" After all, anomalies like ANITA's generate all kinds of interesting new ideas—both theoretical models that no one thought about before and proposals for new signals that experiments could search for. Shoemaker said some dismiss this as "ambulance chasing," but he believes it helps keep physics a thriving and creative field.

Nichol hasn't allowed two curious blips to distract him from his larger aim. He's still interested in the four near-horizon particles seen in ANITA's final flight. It's quite possible that these events are the first evidence of the ultra-high-energy neutrinos that ANITA originally set out to search for.

Nichol is now working on a successor to ANITA, called PUEO, which stands for the "Payload for Ultrahigh Energy Observations." Like ANITA, PUEO will be launched into the Antarctic skies on a NASA balloon. It will be ten times more sensitive than its predecessor. If more strange events are out there waiting to be discovered, PUEO will find them. If confirmed, such particles would allow physicists to probe fundamental physics at energies far higher than we could ever hope to reach at a particle accelerator. Now, that would be one hell of a prize.

Chapter 5

A Magnetic Mystery

*Could the magnetism of a tiny particle called a muon
reveal the hidden influence of new forces?*

Under the cover of darkness in the early hours of June 24, 2013, a giant metal ring, fifty feet across and wrapped tightly in white plastic, squeezed its way out of the main gate of Brookhaven National Laboratory on Long Island, New York, and began a slow procession down the William Floyd Parkway on the back of a specially adapted truck, flanked by a police escort, yellow lights flashing in the dark. On the grassy bank at the road's edge, crowds of curious onlookers gathered, some reclining on lawn chairs as they took in the strange spectacle.

Walking alongside the otherworldly object was Chris Polly, physicist and leader of the project that would ultimately see the ring transported 3,200 miles by land, sea, and river to be installed at the heart of a new experiment at Fermilab near Chicago. Polly had spent the past few years preparing for this moment, working with the transport company to plan the route to Smith Point Marina, where the ring was to be loaded onto a barge; liaising with New York State and Suffolk County Police to close the road section by section; and even arranging to have trees cut down to make sure the ring wouldn't get snagged on a stray branch. It was a delicate procedure; if the fragile ring flexed by more than just a few millimeters during its journey, it could end up broken beyond repair and the entire multimillion-dollar project would be over before it had begun.

Despite the stakes, Polly was relaxed that evening, stopping to chat with the crowds as the ring made its stately progress along the highway. The journey had been planned to the last detail; all that was left now was to enjoy the moment. Plus, it was a great opportunity to share his passion for a scientific project that had consumed almost his entire career. Ten years earlier, the ring had formed the heart of an experiment at Brookhaven that had found tantalizing evidence of hitherto undiscovered ingredients of our universe, lurking unseen in every inch of empty space. If the effect it had detected was real, it promised to transform particle physics and bring us a big step closer to a complete understanding of the most basic building blocks of nature.

Polly had joined the Brookhaven experiment as a young PhD student back in 1997 and had been hooked from the start. The experiment's goal was to measure the magnetism of a fundamental particle called the muon, an exotic, heavier cousin of the more familiar electron. If their measurement differed from the value predicted using the standard model of particle physics, that could indicate that unknown fundamental particles were messing with the muon's magnetism, providing a crucial clue to what deeper theory might one day replace the standard model.

When the experiment produced its first hints in the early years of the twenty-first century that muons weren't behaving as the standard model said they should, excitement had swept through the particle physics community. However, the Brookhaven experiment wasn't precise enough to be sure that the effect they were seeing was real. To be certain the anomaly wasn't either a statistical fluke or a systematic error, Polly realized that they would need a souped-up version. Thus, the Fermilab "Muon $g - 2$ experiment" was born, and Polly was catapulted from newly minted PhD to the project manager of a multimillion-dollar, state-of-the-art particle physics experiment.

However, to make the project affordable, the Brookhaven ring would need to be reused. Hence the herculean logistical challenge of transporting it across Long Island, down the Atlantic coastline, around Florida, through New Orleans, and ultimately up the Mississippi River to its new home outside Chicago.

The ring itself is a marvel. A giant superconducting coil, it generates a powerful magnetic field when an electrical current passes through it. Muons produced by a particle accelerator can be fired around the ring, orbiting in its magnetic field and ultimately allowing physicists to measure their magnetic properties (more on all that in a bit).

However, many of the people Polly spoke to that evening were more than a little skeptical of his explanation of the ring's purpose. Rumors had long swirled around the neighborhood of a UFO that had crashed on Long Island in the 1990s. According to local legend, the putative alien spacecraft had been rapidly whisked away by the white coats from the secretive Brookhaven National Laboratory. And now, lo and behold, a strange saucer wrapped in white plastic emerged from its main gate. One onlooker even stopped Polly with the challenge "You can't tell me that's not a fucking spaceship!"

Polly did his best to disabuse the locals of this notion, but there were some who wouldn't be convinced otherwise. It probably didn't help that the team from the moving company, Emmert International, had stuck a life-sized green alien to the bulge of the ring's cryogenic infrastructure, which looked for all the world like the cockpit. But conspiracy theories aside, the move went off without a hitch, and before the sun had risen, the ring was safely parked at the marina. Later that day, the fifty-ton saucer took flight as a heavy crane hoisted it into the air and ever so carefully lowered it onto the waiting barge. The following day, attached to the plucky little tugboat *Trident*, the ring set to sea on an epic voyage that would take it, Polly, and more than a hundred scientists and engineers from the Muon $g - 2$ experiment into uncharted quantum mechanical waters. Whether they would find monsters or buried treasure, no one could be sure.

The Incredible Weirdness of Muons

At this point, you might well be wondering why Polly and his colleagues went to such enormous lengths just to figure out how magnetic muons are. Hell, you might not unreasonably be wondering what on earth a muon is. If so, you're in good company. Although they were discovered almost a century ago, we still don't have much of a clue of why muons exist. And what's more, they seem to be implicated in several anomalies that are currently shaking the edifice of the standard model. As the cosmologist Will Kinney recently put it, "Muons are weird little fuckers."

The physicist Isidor Rabi would probably have agreed with the sentiment, if not the coarse language. When the muon first appeared on the scene in the mid-1930s, Rabi captured the confusion the new particle was causing with the now famous[*] quip "Who ordered that?"

The muon made itself known to the world by appearing in a cloud chamber that had been lugged to the top of Pikes Peak in the Colorado Rockies by the American physicists Carl Anderson and Seth Neddermeyer. Cloud chambers are almost magical devices that can make subatomic particles visible to the naked eye by causing trails of tiny liquid droplets to form in their wakes, looking like the contrails of a passenger jet seen from a great distance. In 1936, Anderson and Neddermeyer discovered traces left by what appeared to be a peculiar new type of particle that was raining down from the heavens. Perplexingly, these new celestial arrivals didn't appear to play any role in forming atoms, nor did they seem to have anything to do with the strong force that bound the atomic nucleus together, a phenomenon that was only beginning to be understood at the time. In fact, the muon, as it became known, didn't seem to fit in anywhere. It was, to borrow from Rabi's retort, a pizza nobody had ordered.

So, what is a muon? Well, it's a lot like an electron. It is an apparently fundamental particle that has an electric charge of minus one. It feels the pull of the electromagnetic and weak forces—the interactions that give rise to electricity, magnetism, light, and certain types of radioactive decay—but not the strong force, which is responsible for gluing quarks together inside atomic nuclei. What distinguishes the muon from the electron are its mass—two hundred times bigger—and its instability. Left to its own devices, a muon will decay into an electron and a couple of neutrinos after about 2.2 millionths of

* Famous among particle physicists at least.

a second. An electron on the other hand is immortal, living as long as the universe itself, barring accidents.*

Another property that the muon shares with the electron is known as spin. Spin is one of those infuriating quantum mechanical concepts that sounds deceptively familiar but actually isn't at all. When you think of spin, you might think of a coin spinning on its edge, or possibly the thing people who are better at tennis than me do to balls. It's tempting therefore to think of a particle's quantum mechanical spin as the particle rotating on its axis. And indeed, like everyday spin, quantum mechanical spin carries something called angular momentum—a sort of rotational oomph. However, if you try to calculate how fast an electron needs to be rotating in order to account for its measured angular momentum, you discover that it needs to be rotating faster than the speed of light. And as we've now seen, traveling faster than light is a big no-no according to Einstein.

So particles behave as if they were spinning, except they aren't? Don't blame me; blame quantum mechanics.

The differences don't stop there. A tennis ball can have any amount of spin you like, limited, I suppose, only by your backhand slice. But particles are allowed only discrete amounts of spin. Quantum mechanical spin comes in units of ½, with fundamental particles allowed to have spins of only 0, ½, 1, ¾, and 2. In other words, spin is quantized, and electrons and muons both belong to the family of matter particles known formally as fermions, with spin ½.

Why am I banging on about spin so much? Well, this is

---

* Like having an unfortunate run-in with an antiparticle, as we saw in the story of the earliest moments of the universe.

where the magnetism of the muon comes in. As discussed, muons are electrically charged. When you make electric charges move, they generate magnetic fields. This fact was first discovered by the Danish scientist Hans Christian Ørsted in 1820, when he found that a compass needle was deflected when brought close to a wire carrying electricity. The profundity of this discovery cannot be overexaggerated; it was the first step in the unification of electricity and magnetism that would ultimately lead to the standard model. What's more, it spawned the electric motor, the electric generator, and pretty much all electromagnetic technology, which today includes more or less *all* technology. It is also the principle on which Brookhaven's superconducting ring / flying saucer works: a big electric current goes around a big loop of superconducting wire and creates a big magnetic field.

A charged particle with spin behaves like a rotating electric charge (even though it isn't really rotating, sorry again) and therefore generates a magnetic field; in other words, an electron or muon behaves like a tiny electromagnet. This, therefore, is the origin of the muon's magnetism. However, the million-dollar question is, how magnetic is a muon? And moreover, why should anyone care?

To answer these questions, we need to go back to the glory days of theoretical physics, when the ideas that underpin the standard model were born. And before we can really start thinking about muons, we will have to think a little deeper about their close cousins, electrons.

The Invisible Ocean

The question of how magnetic an electron is was first answered by the brilliant British theoretical physicist and quantum superstar, Paul Dirac. In the pantheon of twentieth-century theorists, Dirac is perhaps second only to Einstein. He made foundational contributions to the development of quantum mechanics, though he is perhaps most famous for a single, "achingly beautiful" equation. In 1928, Dirac pulled off a feat that had eluded the greatest minds of the day, the first successful description of the electron that was consistent with both special relativity—Einstein's theory of space and time—and quantum mechanics. Dirac's equation was a mathematical miracle: not only did it match the latest experimental data on how electrons orbit atoms far more precisely than traditional quantum theory, but it also, almost incredibly, predicted the existence of antimatter *before* anyone had seriously imagined that such a thing might exist.

Another of the equation's magic tricks was that it revealed that the electron's spin was an unavoidable consequence of combining special relativity and quantum mechanics; in other words, Dirac could have predicted the existence of spin if it hadn't already been seen in experiments. And most relevant to our story, the equation could also be used to calculate how magnetic the electron should be, given its electric charge, its mass, and in particular how much spin it has.

Now, apologies in advance for descending deep into theory jargon for a second, but this bit is the key to this whole chapter: the relationship between how much magnetism you get for a given amount of spin is encapsulated in a number called the g-factor. The bigger the g-factor, the more magnetic the particle is. Using Dirac's equation, the g-factor for an electron comes out as a perfect, pristine 2. That's just 2, not 2 and a bit, or even 2.000000001. Two exactly. What this means in practice is that an electron is twice as magnetic as it would be if it really was like a tiny spinning (electrically charged) tennis ball, instead of a point-like quantum particle.

The Dirac equation ruled the quantum roost for the best part of two decades. Wielding Dirac's masterpiece, quantum physicists made enormous progress in understanding how light interacted with matter, laying the foundations for the theory that would form the first part of the standard model of particle physics. Within four years of the equation being published, Carl Anderson captured a photograph of the first particle of antimatter—the antielectron—vindicating Dirac's boldest prediction of all.

However, by 1947 cracks were starting to appear. In June of that year, a select group of the world's top physicists gathered on Shelter Island, New York, at the end of Long Island. You'll recall from our discussion of quantum field theory that this was the conference at which the experimentalist Willis Lamb had stolen the show. He had been microwaving hydrogen atoms with Robert Retherford at Columbia University and in the process had discovered a one-in-a-million difference between two energy levels that, according to Dirac's equation, ought to be equal. This tiny anomaly gave theorists the crucial clue needed to properly understand the quantum field theory of light and matter.

Lamb and Retherford's anomaly overshadowed another, equally significant experimental discovery, one made at Columbia at just the same time and also presented on Long Island. Isidor "Who Ordered That?" Rabi had news from his Molecular Beam Laboratory: something was up with the electron's magnetism. Using the phenomena of nuclear magnetic resonance for which Rabi had recently won the Nobel Prize (and which is now the basis on which hospital MRI* machines work), Henry Foley and Polykarp Kusch had found evidence that the electron was ever so slightly more magnetic than predicted by the Dirac equation. Instead of being exactly equal to 2, the electron's g-factor seemed to be around 2.002—a one-in-a-thousand discrepancy that held yet another clue to the true nature of matter.

In the audience at Shelter Island were two ambitious young theorists keen to make their marks on the physics world: Julian Schwinger and Richard Feynman. They were the same age and both from New York City, though the similarities ended there. Schwinger, born to a prosperous family from upper Manhattan, was short, stocky, and urbane; a sharp dresser who prized elegance both in his life and in his physics. By contrast, the tall, lean Feynman hailed from Queens and spoke with such a strong New York accent that some colleagues suspected it was an affectation, part of a carefully cultivated image as a wise-cracking, street-smart, antiauthoritarian "half genius; half buffoon."† At the time, the two men were fighting to tame the troublesome infinities that plagued the nascent theory of light and matter known as quantum electrodynamics. The two men saw in both Lamb's and Rabi's presentations potential keys to the puzzle.

Shortly after the conference, Schwinger got married. He spent his honeymoon pondering the problem of the electron's

---

* MRI stands for "magnetic resonance imaging." Strictly the technique should be called *nuclear* magnetic resonance imaging, but physicians were worried that their patients would be unwilling to climb inside a machine with "nuclear" in its name.

† As the physicist Freeman Dyson described him.

magnetism (a hazard of marrying a physicist). Schwinger wasn't able to get down to work on the problem in earnest until September, but just a couple of months later he had a solution in his hands.

His chance to showcase his work to the world came in January 1948 at a meeting of the American Physical Society in New York. Schwinger presented his calculation of the electron's magnetism to a packed hall, arousing so much interest that he had to repeat it a second time in the afternoon. His calculation caused an immediate sensation, the physicists in the crowd hailing him as a hero. *The New York Times* reported on the excitement, saying "theorists regarded [Schwinger] as the heir apparent to Einstein's mantle and his work on the interaction of energy and matter as the most important development in the last 20 years."

Sitting in the audience was a frustrated Richard Feynman, who had also been working on the electron's magnetism, albeit using a radically different method. At the end of Schwinger's talk, Feynman got up to announce that he had arrived at the same result. However, while Schwinger's reputation as a young man of prodigious talent preceded him (he had got his PhD aged just twenty-one and had just become a full professor at twenty-nine), Feynman was still a relative unknown. He said he ended up feeling like a "little squirt," shouting, "I did it too, Daddy!"

Feynman's chance would soon come. At the end of March, a follow-up to Shelter Island was held at a large hotel in the Pocono Mountains in Pennsylvania. On the second day of the meeting, Schwinger gave a marathon account of his revised theory of quantum electrodynamics, starting in the morning and continuing late into the afternoon. Making his case forensically, step by step, and fielding questions as he went, he eventually built to a dramatic climax: plugging some numbers into an equation, he showed his new theory accurately accounted for the anomaly that Lamb and Retherford had uncovered in their microwave experiments. By the time Schwinger left the stage, the audience was simultaneously impressed and exhausted.

Feynman, normally a supremely confident and charismatic communicator, was a bag of nerves when he stood up to speak next. Not only did he have to follow Schwinger's masterful performance, but several of the world's greatest physicists were in the audience, including his hero, Paul Dirac. Unsurprisingly given the pressure, things did not go well.

Feynman was presenting his own version of quantum electrodynamics for the first time, but one that appeared radically different from Schwinger's. It was full of unfamiliar ideas and methods of his own devising, from particles traveling backward in time to strange hieroglyphic diagrams that he claimed could be used to calculate how particles interacted with one another. When he was asked to justify his methods, Feynman's unsatisfying response was that they gave the right answer. Schwinger later recalled that Feynman's unrehearsed presentation appeared to be "a patchwork of guesses and intuition." Before long the audience grew impatient. Niels Bohr, father of quantum theory, completely derailed Feynman's talk at one point by coming to the blackboard to do his own calculations. Meanwhile, Dirac repeatedly interrupted Feynman with the enigmatic question "Is it unitary?" Feynman had no idea what he was asking, let alone the answer.

He left the stage dejected. His ideas were too unfamiliar and too unformed to be accepted on the basis of a single, rather shambolic talk. However, in time, it would eventually be realized that despite appearing worlds apart, Schwinger's and Feynman's theories were ultimately describing the same fundamental objects, the quantum fields that would come to revolutionize physics. And what's more, it was Feynman's weird pictographs, not Schwinger's elegant mathematics, that would become the primary tools used by particle physicists for decades to come.

You might have noticed that so far I've studiously avoided describing what Schwinger and Feynman had suggested that was so different from Dirac's theory. Well, this was the essence of it: Dirac's theory described the electron as a single, point-like particle sitting in splendid isolation in totally empty space. Schwinger and Feynman both knew that empty space didn't really exist. As we've seen, the quantum field description of nature tells us that particles are not fundamental at all: they are vibrations in quantum fields, invisible but ever-present physical entities that fill every cubic centimeter of space. Thus space, or as physicists call it, the vacuum, is a rather more interesting and dynamic place than in Dirac's theory. As a result, no particle is an island, entire of itself; it sits amid an invisible quantum ocean, constantly nudged and buffeted by its currents and eddies. When viewed from a distance, these currents and eddies in the surrounding ocean are indistinguishable from the pure electron.

In other words, the entity we call an electron is not merely an electron but the combination of the electron *plus* a sort of quantum mechanical froth, made up of wobbles and bubbles in all the quantum fields present in the vacuum.

This means that when you measure the properties of an electron, for instance its magnetism, and you get a slightly different answer from what you'd expect from the Dirac equation, that difference comes from the vacuum itself. What Schwinger and Feynman had done in their own distinct ways was to find a way to calculate the subtle influence of nature's quantum fields on the electron, and hence explain why Rabi's team at Columbia had measured that one-in-a-thousand shift in its magnetism.

The trouble with all of this is that the way an electron interacts with the vacuum is mind-meltingly complicated. For instance, because the electron is electrically charged, it interacts with the electromagnetic field—the field responsible for electric and magnetic forces and light. Thanks to the uncertainty inherent in quantum phenomena, the electromagnetic field is constantly wobbling imperceptibly, and these wobbles push and pull on the electron, altering its properties. However, it doesn't stop there. The electromagnetic field is connected to other quantum fields, for example, the muon field, and so wobbles in the muon field can indirectly affect the electron by setting off wobbles in the electromagnetic field, which then affects the electron again. And so it goes, on and on and on, with every field affecting every other, and all of them ultimately contributing to the electron's properties. It's almost as if the electron sits at the center of a tangled spiderweb, where twanging one strand sets the adjacent strands wobbling too, and eventually the whole web. Today we know of seventeen distinct quantum fields—the strands making up the web if you like—and every one of them messes with the electron's properties. No one has yet managed to figure out how to calculate the total effect of all these wobbles on wobbles on wobbles. Feynman's masterstroke, however, was to devise a wonderfully visual way of breaking the problem down into manageable chunks and, most important, figuring out which bits of this complex picture were the most important and which could be safely ignored. Today, they are known as Feynman diagrams.

As an example of what a Feynman diagram looks like, let's take the way an electron "talks" to the other quantum field in nature, which is primarily via the electromagnetic field. Feyn-

man realized that there was a way to break down this complicated interaction into pieces, with each piece represented by one
of his diagrams. For instance, the simplest way to represent the
electron's interaction with the electromagnetic field is via the
diagram below.

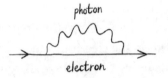

The electron is the straight line with an arrow pointing from
left to right, while the wiggly line is a photon, the particle associated with the electromagnetic field. This diagram appears
to show an electron giving off a photon before reabsorbing it
later—in other words, the electron interacting with the electromagnetic field. Feynman's discovery allowed him to establish a
set of rules to convert each such diagram into a mathematical
expression, one that could be used to calculate that part of the
effect of the electromagnetic field on the electron.

However, there isn't just one diagram representing the electron interacting with the electromagnetic field; there are an
infinite number! To make a more complicated diagram, all you
need to do is add more lines and wiggles. Here are a couple of
examples:

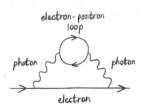

All of these diagrams contribute to the overall effect that you measure in an experiment. This might sound catastrophic, since you obviously can't draw an infinite number of diagrams (you'd run out of paper for one thing). However, the neat bit is that Feynman's rules say that the more complicated a diagram is, the smaller its contribution is. In other words, the more lines and wiggles a diagram contains, the less it matters. That means if you want to get a decent estimate of the effect of the electromagnetic field on the electron, you only need to bother with the simplest diagram. You can then, if you choose to, add more complicated diagrams, which will refine your calculation, bringing it closer and closer to the real-world value; the more diagrams you draw, the more accurate your calculation.

This approach was completely unfamiliar to physicists when Feynman first presented it at the conference in the Poconos, which in part is why he received such a dismissive reception. However, today Feynman diagrams are particle physicists' bread and butter. I regularly find myself doodling a Feynman diagram or two when trying to figure out whether some particle decay might be interesting to look for at the Large Hadron Collider. Meanwhile, theorists can spend years of their careers calculating the next most complicated set of diagrams in a series in order to refine their predictions.

At the Poconos meeting, however, Feynman only needed to worry about the simplest diagram. When he applied his rules and calculated the correction to the electron's magnetism, he got an answer that neatly matched the experimental shift reported by Rabi at Shelter Island, as well as the theoretical result calculated by Schwinger.

All right. Thanks for sticking with me through some rather technical stuff there. Hopefully, it will have been worth it, because we can now, finally, start to understand why measuring the electron's, or for that matter the muon's, magnetism is a very, very, very interesting thing to do. As we've just seen, the electron's magnetism is affected by the vacuum—by all seventeen of the quantum fields described by the standard model of particle physics and which are hanging around unseen in empty space.

However, there are very good reasons to think that there are unknown quantum fields yet to be discovered. For instance, there's dark matter, that as-yet-unknowable stuff that the universe is teeming with. Assuming dark matter is a particle, there will also be a corresponding dark field. If this dark field interacts with any of the particles in the standard model, it will also contribute to the electron's properties. Therefore, if we can measure the electron's magnetism extremely precisely *and* calculate its value in the standard model by drawing lots and lots of Feynman diagrams, and we see a difference between these two numbers, we will have discovered a new field and will have finally broken the half-century hegemony of the standard model.

Indeed, physicists have been improving both the measurement and the prediction of the electron's magnetism for more than seventy years now, reaching a phenomenally impressive precision. The theoretical calculation has grown into an absolute monster. To work out the effect of the electromagnetic and electron fields alone, theorists have calculated contributions from 13,643 different Feynman diagrams. And that's before we even consider contributions from the weak force, strong force, and quarks (the last two of which are especially difficult). For obvious reasons, no one has actually sat down with a pen and paper and drawn all these diagrams by hand. Instead, several teams of theorists have used supercomputers to crunch the numbers, resulting in a state-of-the-art prediction of the electron's magnetism or *g*-factor. It is

$$g = 2.00231930436321$$

That's a calculation to an astonishing fourteen decimal places.* Similarly, the experiments have come by leaps and bounds. The world's most precise measurement of the electron's *g*-factor was carried out by a team from Northwestern and Harvard Universities in 2022. This extraordinary experiment worked by holding a *single* electron in a specially designed trap. The electron was then hit with microwaves to make its spin flip direction. By carefully measuring the frequencies of the microwaves needed to make the electron flip, the Northwestern-Harvard team could measure the *g*-factor. They arrived at

$$g = 2.00231930436118$$

with an uncertainty of just 0.13 parts per trillion! Note the ludicrously good agreement between these two numbers. They are the same until you reach the twelfth digit! (Not only that, but the small discrepancy between them falls relatively comfortably within the theoretical uncertainty.) It's an unequivocal vindication of the standard model and is by far the most accurate agreement between theory and experiment anywhere in science.

On the downside, it's also telling us that, so far at least, there's no evidence that the electron is getting buffeted by quantum fields other than the ones we already know about. But the elec-

---

* Schwinger's original calculation of the *g*-factor was 2.002324—to six decimal places. In seventy years the state-of-the-art prediction has been made a hundred million times more precise.

tron isn't the only game in town. When it comes to the muon, it's a whole different story.

The Saucer Has Landed

I met Chris Polly on a bright July morning at the foot of Wilson Hall, a glass-and-concrete tower that dominates the sprawling Fermilab site. The United States' national particle physics laboratory, Fermilab is set among picturesque prairie land near Batavia, Illinois, on the western outskirts of Chicago. Driving onto the site, I'd passed charming clapboard farmhouses—which I was astonished to discover were provided as homes for visiting researchers—along with herds of bison grazing among the tall grass and joggers out for their morning runs. Standing at the head of a long reflecting pool, Wilson Hall is a magnificent brutalist structure completed in the early 1970s whose tapering concrete flanks and lofty atrium evoke the nave of a cathedral. As a first-time visitor to Fermilab, I was struck by how much thought had clearly gone into the aesthetics of the site, a far cry from the bleak utilitarianism of CERN.

I had come to Fermilab to see the Muon $g - 2$ experiment, which Polly had led from its inception in the middle of the first decade of the twenty-first century. Greeting me with a firm handshake, he led me past security and up into the atrium, a wide open space adorned with greenery, including two full-grown trees. I found my eyes drawn upward by the hall's inwardly curving walls that soared two hundred feet into the air above us. "This is the gorgeous vision of Robert Wilson," he explained, "who felt we should have our own cathedral to science." Wilson was Fermilab's first director, who as well as being an eminent physicist had studied sculpture at the Accademia di

Belle Arti in Florence. He wanted to create a working environment that would elevate the spirits and help draw the best scientists from around the world. A number of his own sculptures can be found across the Fermilab site, including a "hyperbolic obelisk" at the end of the reflecting pool outside the hall.

The building was quiet that morning. I felt particularly grateful to be there when we reached the top-floor viewing deck, which offered breathtaking views of the surrounding countryside. "It does give you vertigo," Polly remarked, smiling as I gawked. Far below us was the reflecting pool, which formed part of a long avenue stretching away through fields and forests to a vanishing point on the far horizon. To the right was a 6.3-kilometer ring-shaped pond encircling a larger irregular lake, the cooling pool of the now defunct Tevatron, the world's mightiest particle collider until it was put out of business by the LHC in 2011.

It had been two hundred feet below us, on the morning of Friday, July 26, 2013, that the Muon $g - 2$ experiment's superconducting ring had been welcomed to Fermilab by a crowd of more than two thousand, who had gathered to celebrate the completion of the "big move." Having braved the Atlantic coastline, Hurricane Alley, and the Gulf of Mexico, the barge had been towed up the Mississippi River until it finally disembarked at Lemont in northeastern Illinois. From there it had been moved gingerly, over the course of three days, through thirty miles of Chicago suburbs, closing two interstates in the process, before crossing onto the Fermilab site just before dawn.

A video recording from the day shows Polly grinning with disbelief that he and the team had managed to bring off such an ambitious plan. "It's amazing," he says in the video, smiling while shaking his head. However, as the crowds dispersed at the end of the day's celebrations, ominous storm clouds rolled in over Wilson Hall. It was a sign, perhaps, that the real challenges were about to begin.

Walking around to the south side of the building, Polly and I paused to chat on a sofa with a bird's-eye view of the building that now houses the experiment. "The nerve-racking part," Polly recalled, "was that the magnet hadn't been turned on for twelve years and we had no way of testing it at Brookhaven. So, we just had to move it and cross our fingers. We were all wondering nervously whether the whole thing was really going to turn on again."

Polly's love affair with experimental physics began as a young undergraduate when he got the chance to spend the summer of 1995 working at the University of Washington. His course hadn't included much fundamental science, but over that summer he got to work on an experiment that was testing the strength of gravity. "I didn't realize people made those precision fundamental measurements," he told me. However, it was catching sight of the "awesome particle accelerator" in the adjacent lab that convinced Polly he wanted to get into particle physics.

He had just gotten married and his wife had landed a dream job, so when looking for a PhD place, he drew a circle around their home and wrote to the best six physics professors in the area. "I got letters back from all six of them, five were short," he recalled. But not David Hertzog's from the University of Illinois. "Hertzog wrote me a three-page letter, saying you should come work with me because the Muon $g - 2$ experiment is the coolest thing in particle physics right now. We're building the calorimeters and you can join us at Brookhaven and shoot a beam into them. And I was like, I'm sold!"

At the University of Illinois, Hertzog was leading work on the first version of the Muon $g - 2$ experiment, then being built at Brookhaven on Long Island. The plan was to measure the muon's g-factor as accurately as possible in the hope of detecting signs of new quantum fields affecting the muon.

We've seen that measurements of the electron's g-factor haven't revealed any signs of hidden fields (so far). But muons have the potential to be even more sensitive to their effects, for the simple reason that they're much more massive than electrons. Being two hundred times heavier means they ought to interact much more strongly with undiscovered fields, whose associated particles are also likely to have large masses.

Put another way, there's more likely to be treasure hidden in the quantum froth that bubbles around the muon than there is around the lowly electron.

As a member of Hertzog's team, Polly was first tasked with figuring out why the calorimeters—detectors that measure particle energies—weren't working as expected. He soon became a core part of the team, spending the next seven years on the project. "I was a long hauler—I could have graduated earlier, but I loved that experiment," he told me. The first really exciting moment came when the team analyzed data from an initial engineering run. To their amazement, the muon's magnetism came out more than three sigma away from the standard model prediction. (As we saw earlier, that means there is a less than one-in-a-thousand chance of getting such a big discrepancy just by dumb luck. Which on our handy sigma scale is referred to as evidence.) With more data on the way, Polly and his colleagues were convinced that the anomaly was on course to cross the all-important five-sigma threshold, where a genuine discovery could be declared.

However, in a dramatic twist that foreshadowed the trouble to come, a mistake was found, not in their experiment, but in the theoretical prediction of the muon's magnetism. In this case, it was the most classic of mathematical cock-ups—one that afflicts school students and Nobel Prize winners alike—a dreaded "sign error" where a plus sign is inadvertently switched for a minus. When the errant minus sign was corrected, the theoretical prediction moved decisively toward the measurement from that first engineering run. The three-sigma deviation evaporated.

And yet, as the experiment recorded more data and the experimental uncertainty shrank, something unexpected began to happen: the value of the muon's magnetism grew less and less consistent with the predicted value. The size of the discrepancy was minuscule, just 2.9 parts per billion, but the uncertainty on the shift was even smaller, a mere 0.8 parts per billion. By the time the Brookhaven team released their final result in 2006, the tension with the standard model had grown to over three sigma.

Could this at long last be the smoking gun that would reveal the presence of particles or forces never seen before? The muon $g - 2$ discrepancy quickly became the most famous anomaly in particle physics, and as the Large Hadron Collider continued to find no evidence of any new particles during the 2010s, for some it became particle physics' last best hope.

However, many remained skeptical. The anomaly was still only hovering just above the three-sigma level. Slaying a theory that had withstood more than fifty years of experimental bombardment would take more than just a piffling one-in-a-thousand wobble. Remember Carl Sagan's caution: extraordinary claims require extraordinary evidence.

After a couple of short postdoctoral positions on neutrino experiments, Polly got together with five others, including his old mentor David Hertzog, to propose a new and improved version of Muon $g - 2$. Their goal was to settle the argument once and for all. The plan was ambitious—a brand-new version of the experiment, where every aspect, from the detectors to the final analysis, would be overhauled and upgraded. The only part that would need to be reused was the superconducting ring from Brookhaven.

After a fierce competition for funding, the project was greenlit and work could begin. At the time, the U.S. Department of Energy was moving away from letting physicists manage large experimental projects, preferring to bring in professional project managers. However, the Fermilab director went to bat for Polly, arguing that the promising young physicist had to be the one to lead the project. So it was that, against the agency's better judgment, Polly found himself heading up a project that would take more than a decade and tens of millions of dollars to deliver.

The arrival of the ring at Fermilab was a welcome boost, both for the collaboration and for Polly personally, proving to the higher-ups that he and the team could pull off a complex logistical move. Of course, it would all be for nothing if they couldn't get it powered up again, something they wouldn't be able to put to the test for another two years. First of all, the experimental hall needed to be built; the lab hadn't wanted to commit to its construction until the ring had made it safely to Fermilab ("in case it sank on the barge or something," Polly said). After that, Emmert International had to come back to Fermilab to slide the ring into its new home, like a giant CD inserted into a player. Then followed another year of work to get it all reassembled and cooled down to its operating temperature of –269 degrees Celsius.

The moment of truth arrived in 2015, just as a review panel had come to Fermilab to assess the project's progress and, crucially, to decide whether to release the funds needed to finish the experiment. While the panel sat deliberating in Wilson Hall, the Muon $g - 2$ team powered up the magnet for the first time in more than a decade. To their dismay, the magnet "quenched" at 40 percent power; in other words, there was an electrical short circuit somewhere. The nightmare scenario was that the problem was buried deep inside the ring itself, but the team knew the most likely place was where external cables were connected. It would take some time to track down the source, so Polly and his colleagues had to fess up to the reviewers that they had run into a problem. But to everyone's relief, the issue was eventually tracked down to the cable connections; the ring itself was mercifully intact.

Funding was approved. It was full steam ahead on the project.

Descending from our lofty vantage point atop Wilson Hall, we jumped into Polly's car for a short drive to the Muon $g - 2$ building to see the experiment itself. On our arrival, Polly first led me up some stairs into a room crisscrossed by pipes and ducts and suffused with the sound of whirring machinery. This, he told me, was a key innovation of the new experiment—a heating, ventilation, and air-conditioning system that keeps the temperature in the experimental hall stable throughout the year. A steady temperature turns out to be critical if you want to measure the muon's magnetism accurately, which requires an exquisitely precise understanding of the magnetic field produced by the ring. Even small changes in the temperature of the room can make the ring expand and contract, messing with the magnetic field. This had been a real problem at Brookhaven. In the upgraded Fermilab, experiment fluctuations were kept to within just one degree Celsius, regardless of what was going on outside, be it a winter snowstorm or a blazing summer afternoon.

Back down the stairs and along a narrow corridor, we entered the experimental hall itself, a large warehouse-like space with a raised metal walkway running around its edge. I was struck by how loud it was; the constant roar of fans made it hard to hear Polly at times. Climbing onto the gantry, we looked down on the heart of the Muon $g - 2$ experiment, the superconducting ring that Polly and his colleagues had gone to such lengths to bring to Fermilab. In truth, the ring itself wasn't visible, buried as it was beneath a thick layer of white insulation that gave it the appearance, I thought, of a giant toilet bowl (I didn't say as much to Polly).

How does Muon $g - 2$ actually work? Well, the experiment really begins several hundred meters outside the hall, where a beam of protons is accelerated and slammed into a metal target, creating a spray of particles called pions. These pions then pass into a triangular-shaped accelerator where they decay into muons, which are then brought into the experimental hall and fired into the Muon $g - 2$ ring.

Once you've got your muons—around ten thousand of them at a time—you have about a tenth of a second to make your measurements before the next pulse of muons arrives. The ring is a giant electromagnet whose powerful magnetic field keeps the muons orbiting in a circle. The magnetic field is created by coils of superconductor—a material that has no electrical resistance—allowing it to carry enormous electrical currents without overheating, and hence generate an incredibly strong magnetic field. However, the ring superconducts only at very low temperatures, just 4.2 degrees above absolute zero, or −269 degrees Celsius, and so is chilled using a combination of liquid nitrogen and liquid helium.

Now—and this is the clever bit—because the muons are magnetic, they try to align themselves with the ring's magnetic field, in the same way that a compass needle aligns with that of the Earth. However, since muons have spin, they also have angular momentum (that rotational oomph we've discussed), and it is a law of physics that angular momentum is always conserved. As a result, the muons *can't* actually line up their internal compass needles with the ring's magnetic field, because that would change their angular momentum. So instead they end up wobbling around it—a phenomenon known as precession.

This all may be a little hard to visualize, but it's the same effect you get when you set a top spinning on a table. Gravity tries to make the spinning top fall over, but conservation of angular momentum prevents this from happening, and so the spinning top stays upright and wobbles around in a circle (it precesses). The same thing happens to muons in the Muon $g - 2$ experiment. The rate at which they wobble around the ring's magnetic field depends on the strength of their own internal bar magnet—or, in other words, the very thing the experiment is trying to measure.

So in short, by measuring the rate of the muon's wobble, you can calculate how magnetic the muon is.

That's fine in theory. The question, though, is *how* can you actually measure how fast the muons wobble? Well, it's possible to figure this out thanks to the fact that muons are unstable, living for only a few millionths of a second on average before decaying into a positron* and two neutrinos. The neutrinos fly off undetected, but the positrons get swept toward the inside of the ring, where they hit calorimeters, whose job is to measure their energies. And it's by studying the energies of these positrons that physicists can figure out the speed of the muon's wobble.

Here's how it works: When a muon decays, the resulting positron flies off in the direction that the muon's bar magnet was pointing. If the muon's bar magnet was pointing in the direc-

* Also known as an antielectron.

tion the muon was traveling, that gives the positron a boost in energy. On the other hand, if the muon's bar magnet was pointing in the opposite direction to the beam, the positron gets emitted backward and ends up with lower energy.

So, if you measure the energies of the positrons that are produced as the muons fly around the ring, you get a characteristic wiggle in their energies as the muons' bar magnets wobble around the ring's magnetic field. Measure the frequency of this wiggle and, hey, presto! You can calculate the magnetism of the muon.

Phew. Okay, thanks for sticking with me through that rather long chain of reasoning. If you didn't quite follow every step, don't worry too much, the key thing to hold on to is that the experiment measures the frequency of the muon's wobble. Or at least that is the *principle* of the experiment. Making it work is a different matter entirely.

Perhaps the most impressive aspect of the experiment is the exquisite precision required to ensure the ring's magnetic field is as uniform as possible. A constant field all around the ring is absolutely essential; any variations would alter the frequency of the measured wobble. Get that wrong and you could easily miss evidence for new particles or, even worse, claim a discovery when it was really a problem with the magnetic field.

"The construction of it is phenomenal," Polly told me as we looked down on the ring from the gantry. "It's like a six-hundred-ton Swiss watch." Once the ring had arrived at Fermilab, sixty-four huge slabs of steel had to be slid into the jaws of the magnet to act as the magnet's poles. "You're putting these huge pieces of steel—meters long—and you're trying to get those in place to a precision of ten microns; that's the thickness of your hair! A piece of steel that size is like a wet blanket when you need ten-micron precision. If you jack them up in one corner, they flop by ten microns somewhere else. So, you have to remove them, make a fine adjustment, and then reinsert them, remeasure them. It was very laborious."

Of course, even once an experiment has been painstakingly constructed, things never work as they're supposed to straight out of the box. One of the most dramatic teething problems that the team had to solve had to do with a special device called a kicker. The kicker's job is to give the muons a short, sharp shove after they first enter the ring, to stop them traveling around in a perfect circle and smacking straight back into the place they entered by. This shove is delivered by a 160,000-volt electric field. But because it takes the muons only 150 billionths of a second to orbit the ring, this field has to switch on and off in less than 100 billionths of a second. The result was "sparks galore," Mark Lancaster, former spokesperson of Muon $g - 2$, had told me a few months before my visit to Fermilab. The kicker proved so violent that you could hear loud bangs caused by electrical sparks when the experiment was running, even from the safety of the control room. When Lancaster and his colleagues entered the experimental hall after a run, they would often find broken electrical cables and scorching on the equipment.

The first run, in 2018, was "plagued with difficulties," Polly said. "As soon as we turned on the experiment, we found every other thing that needed fixing—so run 1 was not a good year for us." The problem with the kicker meant the muons couldn't be shoved hard enough to get them right into the center of the ring, where their orbits would be stable, which made analyzing the data afterward fiendishly difficult. However, despite the messiness of the data from that first year, it would end up being used for the experiment's very first measurement of the muon's magnetism.

As Polly put it, that first measurement would prove to be their "real Mars landing moment." The question was, where would their ship land? Would the new and improved experiment reveal that Chris and his colleagues at Brookhaven had messed up back in the early 2000s and that any hints of new physics were simply a mirage? If that proved true, not only would it be a personal blow for Polly and his former Brookhaven colleagues, but it could inflict a mortal wound on particle physics as a whole, killing off one of the last hopes for spying new particles. If the anomaly was confirmed, a new dawn could break for our understanding of the universe's most fundamental building blocks. The scientific stakes could hardly have been higher.

Unblinding

On a cold, crisp February afternoon in 2021, Chris Polly walked alone across the echoing atrium of Wilson Hall clutching a sealed envelope. Inside was the key to the mystery of the muon's magnetism and with it the careers and reputations of hundreds of scientists, the future of an entire field, and perhaps the universe's deepest secrets. Thanks to the COVID-19 pandemic, Polly had the towering building all to himself that day. It was a strange culmination for what had been a huge team effort involving more than two hundred people working tirelessly for years to reach this moment of truth. By rights Polly should have been surrounded by his colleagues. Instead, they were waiting with bated breath on Zoom.

Three years had passed since the troubled first experimental run. Since then, the physicists of Muon $g-2$ had been developing the techniques needed to analyze the data and ultimately measure the muon's magnetism, a task made all the more complex by the need to account and correct for the problems of that year of data taking. However, after years of work, discussion, and detailed scrutiny, the team had finally satisfied themselves that their method was solid. Every source of bias that anyone could think of had been considered and estimated. Now all that remained was to open the box and see what nature had in store.

Until that day, the analysis had been performed "blind," which meant that the value of the muon's magnetism had been kept hidden from everyone on the experimental team, Polly included. Blinding is a widespread practice in physics these days as a defense against conscious or unconscious biases creeping into a measurement. The risks of being able to continually look at the answer while you are adjusting your analysis methods are self-evident. Even the most dispassionate scientist has their prejudices and preconceptions, their hopes and fears. Imagine you had worked on both the Brookhaven and the Fermilab experiments; you might well hope that the new Fermilab measurement would confirm the result you got at Brookhaven, and so, when choosing how to analyze the data, you might be drawn toward techniques that move the new measurement closer to the old one. By blinding yourself, the temptation to nudge the result in your preferred direction is removed.

By now, however, the analysis methods had been frozen. The collaboration had voted unanimously: it was time for their measurement to be "unblinded."

As we saw earlier, the experiment determined the muon's magnetism by measuring the frequency with which muons wobbled around a powerful magnetic field. This frequency is simply how many wobbles the muons perform in a given period of time, as measured by the experiment's clock. This clock isn't like the one you might have hanging on your wall at home. This clock ticks around forty million times per second, and the rate at which it ticks can be adjusted within a fairly wide range around this value. If you don't know how fast it's ticking, then you also, deliberately, can't tell how fast the muons are wobbling.

Three years earlier, two members of Fermilab's directorate, Joe Lykken and Greg Bock, were asked by the Muon $g-2$ team to be their "clock people." Their job was to unlock a cabinet in the experimental hall and set the clock frequency to a random value, which they then wrote down on two pieces of paper and placed in sealed envelopes. Crucially, neither Lykken nor Bock is a member of the Muon $g-2$ team. They would keep the frequency a closely guarded secret until the moment came to unblind. ("My wife won't pick me for responsible jobs like this, so I don't know why an important experiment did," Lykken later joked.)

One envelope was kept locked away at Fermilab. The other was mailed to the University of Washington—"just in case Wilson Hall burned down," Polly said. Losing the frequency would spell disaster, rendering all the data recorded by the experiment useless. Such a calamity almost befell the Brookhaven experiment, when the person who was guarding the clock frequency retired and his office was unexpectedly cleared out. In desperation, the Brookhaven team had had to go rooting through a dumpster in search of the crucial scrap of paper. Not a good day at the office.

Sitting alone in the experiment's off-site control room on the ground floor of Wilson Hall, Polly connected to Zoom. He could see just under two hundred faces staring eagerly back. With cold winter sunshine streaming through the window behind him, he tore open the envelope and held up the eight-digit clock frequency for all to see. At the University of Washington, his old PhD supervisor, David Hertzog, did the same with his envelope. The two numbers matched; so far, so good.

Shared on the screen was a digital workbook that would compute the final result once the clock frequency was typed in. Already on display was the old Brookhaven measurement, alongside the state-of-the-art theoretical prediction of the muon's magnetism. Where would the new result land? Would it deviate from theory and confirm the hints of new physics, or would it fall in line with the standard model? As Polly later told me, "Some of my collaborators thought if it didn't agree with Brookhaven, it just meant those Brookhaven guys did something wrong. But I was one of those Brookhaven guys! And I was as confident in that group and their abilities as I am in this group and their abilities."

The clock frequency was typed in. Boom. The new result landed right next to the one from Brookhaven. Spontaneous cheering and whooping erupted across the Zoom feed.

"Wow," Polly said to himself, sinking back into his chair with a mixture of relief and amazement. Like everyone, his eyes had alighted on a crucial number calculated automatically by the workbook: the combined effect of the Brookhaven and Fermilab results. The discrepancy with the standard model had swelled from 3 sigma to 4.2 sigma, meaning there was now a less than one-in-forty-thousand chance of the two numbers disagreeing by dumb luck.

At last, fifteen years after Brookhaven released their final result, the hints of new physics had been confirmed. They had taken an impressive bound closer to that vaunted five-sigma mark. Polly and his team knew that when they revealed their result to the world in just a few weeks, it would cause a scientific earthquake.

A Sting in the Muon's Tail

All good stories have a dramatic twist just when things are looking up for the heroes—the Red Wedding in *Game of Thrones,* the Death Star getting blown up in *Star Wars* (I always find myself rooting for Vader and his pals). Alas, this one was no exception.

Following the dramatic unblinding in February 2021, the Muon g – 2 collaborators began preparations for their grand reveal. A press release was drafted and a special online seminar was organized, wherein the experimental team would show off their new measurement alongside a global consortium of theoretical physicists known as the Theory Initiative, who would present their state-of-the-art prediction using the standard model. The date was set for April 7.

However, just a few weeks before the big day, some unexpected and unwelcome news arrived from a rival group of theorists. They had performed their own calculation of the muon's magnetism using different methods from the Theory Initiative's, and it agreed much more closely with the experiment's result. If their calculation was correct, then the anomaly was no anomaly at all, but another spectacular vindication of the infuriatingly indestructible standard model. All dreams of new particles and forces would evaporate, and physicists would be left with scant clues as to where to go next. To add insult to injury, their journal was planning to publish the paper on the same day as the Muon g – 2's presentation. (Few on the experiment bought that the timing of the publication was a coincidence, despite denials from the theory group that their timing was deliberately designed to steal their experiment's thunder. "Yeah, bullshit," was one Muon g – 2 physicist's reaction.)

Who were these villainous theorists trying to spoil our dramatic story of discovery? They go by the name of the BMW collaboration, not because they drive around dispensing troubling theoretical predictions from sinister black German saloons with tinted windows, but because they are from Budapest, Marseille, and Wuppertal. And of course, unwelcome as their work might have been for those of us hoping desperately for signs of new physics, they're not really villainous (at least as far as I know).

You might reasonably be wondering how it is possible to arrive at two different theoretical predictions of the same quantity using the same theory. Well, the answer is that doing calculations using the standard model is *hard,* and theorists often have to make different assumptions or take different approaches to attack a calculation. Earlier we saw how Richard Feynman came up with a neat shortcut to calculating particle interactions using his handy diagrams—combinations of lines and wiggles representing the electron emitting and reabsorbing photons. We also saw that the state-of-the-art calculation of the electron's magnetism involved calculating contributions from 13,643 different Feynman diagrams, each one helping to account for the effect of the electromagnetic field on the electron's magnetism. Now, you might think that calculating contributions from thirteen-thousand-odd diagrams is hard enough, but actually that's the easy part ("easy" being a relative term). There is another contribution, particularly important when calculating the muon's magnetism, for which this whole Feynman diagram business breaks down. It comes from the most theoretically challenging ingredient of the standard model: the strong force.

To recap, there are three fundamental forces in the standard model: the electromagnetic force, the weak force, and the strong force. The strong force is responsible for holding protons and neutrons together inside atomic nuclei and, more fundamentally, binding quarks together to make protons and neutrons. Now, while muons can't interact with the strong force directly, they can be affected by it indirectly. The electromagnetic force allows a muon to push and pull on the quark fields, which in turn are affected by the strong force. This means that quantum wobbles in the quark and strong force fields form part of the quantum haze around the muon and contribute to its magnetism.

The problem is that our usual approach of drawing Feynman diagrams to calculate the size of the effect doesn't work for the strong force. Whereas for the electromagnetic force you can get away with drawing a finite number of the simplest diagrams while disregarding the infinite number of more complicated ones, for the strong force this trick won't get you anywhere. Because the strong force is, well, strong, making a diagram more complicated by, say, adding an extra gluon (the particle of the strong force) to it doesn't necessarily make the contribution from that diagram smaller. As a result, we have no idea when to stop drawing diagrams. Perhaps the simplest one is the biggest, but it could also be the millionth diagram that matters most. The only way to know which diagrams are important and which aren't is to draw and calculate an infinite number of them, which is, naturally, impossible.

What then are we to do? Well, there are two different ways to get around this problem. The first is the one used by the Theory Initiative—around 170 theorists from around the world who got together to agree on a consensus prediction for the muon's magnetism (the one that disagreed with Muon $g - 2$'s measurement). Rather than try to calculate the effect of the quantum cloud of quarks and gluons that swirls around the muon, they used real experimental data to estimate it. This data came from particle colliders that smash electrons and antielectrons into each other, producing particles made from quarks and gluons.

Because electrons and muons are almost identical, the way an electron collider makes quarks and gluons is very closely related to how those same quarks and gluons affect the properties of the muon. So, theorists can take measurements from electron-antielectron collisions (usually from multiple different experiments) and translate them into a calculation of the contribution of quarks and gluons to the muon's magnetism. The clever thing about this approach is that you avoid having to deal with the headache-inducingly hard theory of the strong force and can instead take the effect directly from the real world. This data-driven method has been perfected over many years, and so as long as the experimental data is solid, so should the prediction of the muon's magnetism.

However, there is a way to calculate the contribution of quarks and gluons to the muon's magnetism *purely* from theory. Instead of using Feynman diagrams or experimental data, the BMW collaboration attacks the thorny problem of the strong force using a technique known as lattice QCD. In essence, this involves breaking space and time down into a grid of discrete points (a lattice) and solving the equations of the strong force only at these points. You can think of this as a bit like taking a digital photograph. Instead of trying to capture an image in all its infinite glory (be it a moody Scottish Highland landscape or your best Blue Steel selfie), a digital camera breaks the image down into a finite number of pixels. As long as the number of pixels is sufficiently high (I'm old enough to remember the days when we used to get excited about how many megapixels our camera phones had), the photo will look sharp and crisp.

Lattice calculations work in a similar way. By breaking space and time down into a lattice, you simplify the problem and make the equations of the strong force solvable. The trick then is to make the lattice as fine grained as possible so you properly capture the key effects while not making the calculation insuperably difficult. This technique had been steadily improving over a number of years, and the BMW team made several significant advances that ought to make their calculation more accurate. However, the downside of their improved method is that the calculation becomes enormously computer intensive, meaning the team had to wangle time on supercomputers in Jülich, Munich, Stuttgart, Orsay, Rome, Wuppertal, and Budapest to run the numbers.

It took hundreds of millions of computer core hours, but when the calculation was complete and added to the prediction of the muon's magnetism, it came out much closer to the Brookhaven result than the "accepted" value calculated using experimental input. If BMW was right, then on the face of it there was no evidence for unknown particles in the muon's quantum mechanical wake and experimenters have spent decades chasing a phantom.

So, it was under something of a cloud of uncertainty that the Muon g − 2 experiment finally announced their brand-spanking-new result. Nonetheless, the news that they had confirmed the Brookhaven result and strengthened the case for new physics caused an international sensation, getting write-ups across the world, including a rather hyperbolic headline in *The New York Times* claiming that the laws of physics themselves had been "upended."

Hot on the heels of the worldwide media blizzard came an avalanche of theoretical papers attempting to explain the origin of the anomaly in terms of exotic new types of matter and energy. Breathless talk of an exciting new era in our understanding of the universe abounded.

But once the dust had settled, the uncomfortable fact of the BMW prediction remained lurking in the background. It is now clear that the mystery of the muon's magnetism will be resolved not by a battle between theory and experiment but by a battle within theory itself.

An Uncertain Future

As Polly and I left the Muon $g - 2$ experimental hall, we stopped by the control room for one final task. "You're witnessing a hallowed and sacred tradition," he told me as a colleague passed him three sealed envelopes. Just days before my visit, the experiment had completed its fifth, and possibly final, experimental run, and in those envelopes was the secret eight-digit clock frequency needed to unblind the previous two years' worth of precious data. Polly's next stop was Wilson Hall, where two of the envelopes would be handed over to the administrators and directorate for safe keeping, while the third would, again, be posted to the University of Washington.

Those envelopes will remain sealed until 2024, when it is hoped the final analysis of the data taken in the fourth and fifth runs will be unblinded. So far, the experiment has analyzed only around 6 percent of its total data set, meaning that its measurement of the muon's magnetism will soon become far, far more precise. If the value stays where it is, then the tension with the Theory Initiative's standard model prediction will easily go over five sigma, crossing the threshold where, at long last, physicists can declare the discovery of new quantum fields.

Of course, that assumes that the Theory Initiative's prediction is correct. If, on the other hand, the new supercomputer calculation made by the BMW team is right, then the picture will be a lot less decisive. In some sense, the ultimate result that the experiment arrives at has become a bit of a sideshow, secondary to the battle over the theoretical prediction of the muon's magnetism. Until that thorny issue is resolved, it will be impossible to make a clear statement for or against new ingredients of our universe.

To try to get my head around this rather puzzling state of affairs, I spoke to Christine Davies, a professor of theoretical physics at the University of Glasgow and an expert on lattice QCD, the technique used by the BMW team to make their prediction. Davies first became interested in the muon's magnetism in 2014 and has since become part of the global Theory Initiative team who produced the "standard" prediction—the one that disagrees with the experimental measurement. However, despite being on the rival team, as it were, Davies expressed confidence in the BMW team's calculation. "I think the lattice results are pretty solid."

Since the BMW team released their prediction in April 2021, several other lattice theory groups have cross-checked their results, and despite each group using different approaches, they all appear to line up rather nicely with the BMW value. As we saw, the lattice technique allows physicists to calculate the effect of quarks and gluons on the muon's magnetism directly from the standard model, while the Theory Initiative uses real-world data from electron-positron colliders to get at the same quantity. "There's a tension developing between the electron-positron results and the lattice results," Davies told me, "and we're trying to understand that tension at the moment. How that story will develop is very unclear."

As Davies sees it, there are two possible explanations for what's going on. "Either there's some bias in the electron-positron collider experiments that's somehow been missed, or

there is new physics in the electron-positron collisions themselves!" The second possibility would be a huge surprise. Bashing electrons and positrons into each other to make quarks and
gluons is regarded as one of the best understood areas of particle physics, and certainly not somewhere you'd expect to see
the influence of hidden quantum fields.

Of course, it seemed to me that there was an obvious third
possibility—that the lattice QCD calculations were wrong.

"Well, yes, I suppose you can't discount that," she said. "It
would be easy for me to go, 'Oh yes, of course there could be
something wrong with the experiments,' but you know as well
as I do how much effort these experiments go to chase down
errors. There are other, more accurate experiments going on,
so we will get more data. They could find a problem with the
old results."

Unraveling this enigma will ultimately require a multi-
pronged approach. Lattice theorists will continue to refine their
techniques and improve the precision of their calculations,
cross-checking each other and hopefully building confidence
in their predictions. Meanwhile, experimenters will be picking over the electron-positron collision data with a fine-tooth
comb in search of hidden errors. In the longer term, brand-new
electron-positron collision data will be coming down the track,
allowing these measurements to be validated and improved.

It's impossible to say where all this will lead right now.
Davies told me her gut is that some mistake will be found in
the experimental data, or perhaps how it's being interpreted by
theorists. "The trouble is that history has told us that new physics is very, very hard to find. I think there's still a lot of shaking down to be done. It wouldn't surprise me if the next Muon
$g - 2$ measurement gets closer to the theory prediction. Things
could start to move in a direction where we end up somewhere

in the middle. But it will take a long time to get there. That's where I would put my money.

"That isn't what I *want* to happen," she added as an afterthought. "We desperately want new physics!"

Indeed, we do. Davies's skepticism is wise, though; the standard model has proved incredibly resilient, even as anomalies have come and gone. If it does turn out that the mystery of the muon's magnetism is down to a misunderstanding of the physics of quarks and gluons, we will at the very least have learned something deep about the strong force. While perhaps less exciting than discovering new particles or forces, it would nonetheless mark a significant step forward in our understanding of the most fundamental ingredients of our universe, and could well lay the groundwork necessary to make future discoveries.

But there are many in the theory community who are far more bullish about the prospects for the Muon $g - 2$ anomaly. And what's more, they've found it possible to connect it to some of the deepest mysteries facing science today—in particular, to dark matter, the invisible stuff that is presumed to be five times more abundant than all the atomic matter in the universe. Through Muon $g - 2$'s findings, light could be shed too on "dark forces," the dark matter versions of the forces we know and experience. We might find that these forces allow dark matter to form structures, perhaps even akin to stars or planets.

Just imagine what that would mean: parallel dark galaxies populated by billions of dark stars, living alongside our own: invisible, untouchable, and just out of reach.

Chapter 6

Ghosts in the Machine

*Could a twenty-year-old mystery around the unexpected
appearance of neutrinos hold clues to a hidden universe?*

They are invisible sunshine, an unseen rain, ghostly travelers from the beginning of time on the way to eternity. They slice through planets and pierce stars. To them our physical universe is insubstantial, a spectacle to which they are almost entirely indifferent. Neutrinos are the strangest and yet most ubiquitous of all the things that make up the cosmos. They are also wrapped up in some of its deepest mysteries.

For every atom in the observable universe, there are at least a billion neutrinos. A hundred trillion pass through you every second. Many were made in the nuclear reactions that power the Sun, others by cosmic rays slamming into the upper atmosphere, some by radioactive elements buried deep underground. But the majority were created in the first moments of the big bang, and they've been streaming through space ever since. Even the bananas in your fruit bowl are giving off a steady trickle due to small quantities of a radioactive form of potassium. Yet we are totally oblivious to this relentless flood since neutrinos hardly ever make contact with ordinary matter. Lacking either electric charge or the charge of the strong force, they can influence our physical world only through the triflingly feeble weak force. As we saw when we met the Antarctic ANITA experiment, this allows them to fly through the Earth unimpeded. It also makes them maddeningly difficult to detect. That's why, despite their abundance, neutrinos are the fundamental particles about which we know the least.

Hunting neutrinos is an obsession for many physicists, and for good reason. While their habit of escaping detection makes them exceptionally tricky to study, it also makes them unique probes of some of the most inaccessible phenomena in nature. They have been used to peer into the heart of the Sun and study the violent deaths of stars, and one day they may even allow us to look back to the very first moments of the big bang. Perhaps most tantalizing of all, neutrinos might have been responsible for creating the slight imbalance of matter over antimatter that emerged during the birth of the universe, ultimately giving rise to everything in existence.

For the past twenty years, neutrinos have also been the cause of a number of unexpected anomalies that some physicists think could presage a transformation in our understanding of nature. Two separate experiments, one at Los Alamos in New Mexico, another at Fermilab, have found evidence of neutrinos appearing in places they have no right to be, according to the standard model of particle physics. Other experiments that study neutrinos emitted by nuclear reactors or radioactive isotopes have also found hints that neutrinos are not behaving as expected. For many years, theorists suspected the anomalies could be evidence of a new type of *even more* antisocial neutrino, one that doesn't even interact through the weak force. Such a discovery would in itself represent a major step forward in our understanding of nature's fundamental building blocks.

But more recently, an even more exciting explanation has emerged. Some theorists have suggested that neutrinos might be affected by dark matter and dark forces. It's a thrilling possibility, but as with all anomalies the picture is murky and uncertain.

From the very beginning, our understanding of neutrinos has been driven forward by anomalies. In the latter half of the twentieth century, neutrino experiments revealed a number of mysterious phenomena that took physicists decades to decipher. However, when their solutions were found, they opened a brand-new route to probe nature's most basic building blocks and led to the first and only revision of the standard model. The story of these earlier anomalies is intimately bound up with the new results surrounding these bizarre particles.

So, to begin, we'll take a short journey back in time to the neutrino experiment where everything started, an experiment whose results were so hard to fathom that it made physicists worry that the Sun might be dying.

The Day the Sun Went Out

I had a dream, which was not all a dream.
The bright sun was extinguish'd, and the stars
Did wander darkling in the eternal space,
Rayless, and pathless, and the icy earth
Swung blind and blackening in the moonless air;
Morn came and went—and came, and brought no day,
And men forgot their passions in the dread
Of this their desolation; . . .

(Lord Byron, "Darkness")

In 1816, panic swept through Europe. According to an astronomer from Bologna, the Sun was about to go out, dooming humankind to a terrifying end as the Earth slowly froze in a never-ending night. The astronomer even obligingly provided a date for the coming apocalypse: July 18. Such doom-laden prognostications might not have attracted much attention in normal times, but 1816 was no ordinary year. Europe and North America had suffered the coldest summer on record, widespread crop failures had led to famine and typhus outbreaks, birds fell dead from the sky, and the air was filled by a strange haze that produced ominous, blood-red sunsets. At a villa on Lake Geneva, Lord Byron and his artistic friends retreated indoors as torrential rain and thunderstorms rolled over the lake, inspiring Mary Shelley to write *Frankenstein* while Bryon penned the poem "Darkness," imagining the grim fate of humanity after the death of the Sun.

Fortunately, July 18, 1816, came and went and the Sun continued to shine. It was eventually realized that the Year Without Summer, as it became known, had been caused by a vast cloud of volcanic material thrown into the atmosphere by the cataclysmic eruption of Mount Tambora in Indonesia.

Around a century and a half later, astronomers were once again worried that the Sun was about to go out. This time the culprits were neutrinos.

Almost a mile down an old gold mine in South Dakota, the physicists Ray Davis and John Bahcall were leading an experiment designed to look directly into the solar core. By 1966, after a decade of arm-twisting and fundraising, they had managed to excavate a cavern in the Homestake gold mine large enough to house a 400,000-liter tank of dry-cleaning fluid, attached to a set of pumps and condensers. Their hope was that this rather peculiar apparatus would allow them to capture neutrinos produced in the Sun's nuclear furnace, directly testing the theory that stars are powered by nuclear reactions that fuse hydrogen into helium.

Chasing neutrinos had been a lifelong obsession for Davis. In the 1950s, he had conducted an experiment at a nuclear reactor at the Savannah River Site in South Carolina in the hope of spying them for the very first time. However, he had been beaten to the finish line by Fred Reines and Clyde Cowan, whose wonderfully named Project Poltergeist was the first to detect neutrinos directly, particles that at one time had been thought to be so elusive that they would never be found.

This time around, as they investigated the Sun's nuclear engine, Davis and Bahcall had the field to themselves. The question was, would their madcap scheme actually work?

The idea behind the experiment was this: if the Sun really was powered by nuclear fusion, then the associated nuclear reactions should produce a steady flood of neutrinos, which would zip out of the Sun at the speed of light and arrive at Earth a short eight minutes and twenty seconds later. Some of these neutrinos would be bound to pass through Davis and Bahcall's giant tank of dry-cleaning fluid, and a tiny, tiny fraction of these would collide with a nucleus of chlorine, converting it into a radioactive form of the gas argon. Helium would then be bubbled through the tank, in order to extract any argon atoms that had been produced, which would then be fed into a special device where they could be counted. If the number of argon atoms matched Bahcall's calculations based on rate of reactions in the heart of the Sun, then they would have conclusive evidence that the Sun really was a nuclear forge.

Bahcall's estimates suggested that solar neutrinos would produce only a few tens of argon atoms each month, a pifflingly small number. To have any hope of spying such a rare signal, Davis and Bahcall had to work meticulously to remove any possible sources of contamination. One obvious problem is that ordinary air is around 1 percent argon. So they went to enormous lengths to purge all possible traces of it from their tank. Another potential spoiler were cosmic rays, those charged particles that rain down from outer space and could easily spoof a neutrino. Interference from cosmic rays had put paid to Davis's earlier attempt to discover neutrinos, which is why he and Bahcall had gone to such expense to bury their experiment beneath a mile-thick shield of solid rock.

By 1968, Davis and Bahcall were ready to reveal their first results. To their delight, the experiment had worked: they had detected a significant number of argon atoms, showing, apparently conclusively, that they were capturing neutrinos from the Sun.

However, puzzlingly, they'd spied far fewer neutrinos than Bahcall had predicted. The shortfall made many suspect there was a problem with their setup. Far from being lauded as the scientists who had finally answered the millennia-old question "Why does the Sun shine?" their discovery was largely dismissed.

Among the doubters was Willy Fowler, a pioneer of experimental nuclear astrophysics who challenged Davis and Bahcall to inject five hundred atoms of argon into the tank, stir them around a bit, and then prove that they could recover them all. Davis and Bahcall duly took up Fowler's challenge. They passed with flying colors, recovering every one of the atoms they'd mixed into the tank, a truly mind-boggling feat. And yet, even after various improvements to their setup, they continued to report seeing solar neutrinos at only around a third of the rate predicted based on the state-of-the-art understanding of the Sun.

By the early years of the 1970s, the "solar neutrino problem" was becoming serious. Suspicion began to fall on the Sun itself, or at least our ability to accurately model what was going on inside it. Particle physicists assumed that solar physicists must have messed up their calculations, because even a small mis-estimation of the temperature of the solar core could result in a large change in the predicted rate of nuclear reactions. However, increasingly sophisticated experiments that re-created nuclear reactions in the laboratory gradually built confidence in the solar model that Bahcall had used.

As the evidence mounted, some physicists alighted on another, more alarming hypothesis: perhaps the Sun itself was broken.

You might think that if the nuclear engine that powers the Sun were to suddenly stop working, we'd notice. But surprisingly, such a disaster might not be immediately obvious. It takes tens of thousands of years for light generated at the solar core to fight its way to the surface. Meaning that if the Sun ran out of fuel and its collapse became a certainty, we could be blissfully unaware of our impending doom for millennia. Maybe, physicists theorized, Davis and Bahcall's experiment was observing fewer neutrinos than expected because the Sun was dying before our eyes.

What now seem like wild ideas started to appear in the scientific literature. In 1973, the Australian mathematician Andrew Prentice suggested that the Sun had simply exhausted its hydrogen fuel supply, leaving a core of inert helium. A couple of years later, a group of theorists speculated that a black hole had settled at the center of the Sun and was slowly devouring it from the inside out.

Despite the apocalyptic consequences of these ideas, the general public remained largely unfazed. Perhaps they had more pressing matters to worry about, like nuclear war with the Soviet Union or the moral panic caused by David Bowie's tight space trousers.

In 1978, physicists got together at Brookhaven on Long Island, New York, to discuss this worrying state of affairs.

Davis and Bahcall had now been reporting the same results for a decade, and every attempt to find the cause of their anomaly—a mistake either in their experiment or in the theoretical predictions—seemed only to confirm the seriousness of the problem. The Sun was now increasingly regarded as the prime suspect. Hardly anyone imagined that neutrinos themselves might be responsible.

Other experiments were clearly needed to cross-check the Homestake results. But what should this next-generation neutrino detector look like?

A big drawback of Davis and Bahcall's dry-cleaning fluid method was that it was sensitive only to the highest-energy neutrinos from the Sun, while the vast majority produced by fusion of hydrogen into helium didn't have enough energy to turn a chlorine nucleus into argon, making the Homestake experiment blind to them. To detect the much more common low-energy neutrinos, a detector would need to be made from gallium, a soft silvery metal prized for its use in electronics. Unfortunately, the proposed experiment would require three times the world's annual output of the element, making it ludicrously expensive. Despite several attempts to get the project off the ground, it never secured funding in the United States. In the meantime, however, a new type of detector technology was emerging that would come to the Sun's rescue.

Deep under the Okuhida mountains in Japan, physicists had installed a giant cylindrical tank containing three thousand tons of ultrapure water. Lining the walls of the tank were a thousand golden orbs—technically known as photomultiplier tubes—that watched for any signs of light emerging from the dark water. The original purpose of the experiment was to search for evidence that protons eventually decay into lighter particles, a key prediction of the then-fashionable "grand unified theories," which attempted to unify the electromagnetic, weak, and strong forces. By 1985, the Kamiokande experiment, as it was known, hadn't seen any signs of protons snuffing it. But the team realized that with a few upgrades it could be repurposed into a brand-new type of instrument: a neutrino telescope.

The idea was this: On the very rare occasion that a neutrino bumped into a water molecule in the tank, it would convert into an electron, which would continue zipping through the water in the same direction as the original neutrino. Now, because the speed of light in water is slower than the speed of light in a vacuum, it's actually possible that the electron could be produced traveling faster than *the speed of light in water,* and would therefore emit a cone of light known as Cherenkov radiation—the light equivalent of the sonic boom emitted by an aircraft when it breaks the sound barrier. This cone of light would be emitted along the direction the electron was traveling, tracing out a ring on the detector-lined walls of the tank, allowing the Kamiokande experiment to detect not just the neutrino but the direction it came from. In other words, Kamiokande would be able to tell whether a neutrino came from the direction of the Sun or indeed somewhere else.

After spending a couple of years upgrading their experiment to improve its sensitivity to neutrinos, the team in Japan began to detect solar neutrinos. But once again, they found significantly fewer than predicted. After twenty years, Davis and Bahcall's claim had finally been vindicated.

More was to come. In 1989, having consumed the lion's share of the global gallium supply, the Soviet-American Gallium Experiment, or SAGE, began searching for lower-energy neutrinos under the Caucasus Mountains in Russia. It was soon joined by the GALLEX experiment under the Gran Sasso mountain in Italy. In each instance, as the data accumulated, it became clear that neutrinos were missing not just at higher energies but at lower ones too.

Meanwhile, increasingly accurate studies of the Sun using helioseismology—essentially measurements of tremors in the body of the Sun, known as sunquakes—confirmed solar physicists' calculations of the conditions in the solar core. They indicated the Sun was in perfectly good health.

If all was well with the Sun, then how to explain the missing neutrinos?

In fact, progress in particle physics had already been hinting at another explanation for the solar neutrinos' absence. You'll recall that back in 1936, Carl Anderson and Seth Neddermeyer had hauled their cloud chamber to the top of Pikes Peak in Colorado and in the process discovered the muon. The arrival of the muon on the subatomic scene had baffled physicists at first, because they didn't know how to fit it into the subatomic bestiary. One clue later emerged from the fact that muons decayed into electrons after about two millionths of a second. It was natural to assume that these new particles were really just overexcited electrons that de-excited back into ordinary electrons, thereby giving off a photon. However, in 1948, Jack Steinberger carried out an experiment showing that muons decayed into an electron and *two* invisible particles, not one.

Over time it became increasingly clear that these two particles were none other than neutrinos. But a mystery remained: Putting aside the neutrinos for a moment, why had no one ever *seen* a muon turn into an electron and a photon?

It was the Italian theorist Bruno Pontecorvo who hit upon a solution. What if electrons and muons carried some kind of essential "electronness" and "muonness" that was always conserved? If that were true, then it would be impossible for a muon to decay into an electron and a photon. *Something* needed to carry off the muon's "muonness" to keep things balanced. Perhaps, Pontecorvo speculated, one of the two neutrinos released when a muon decays carried off its essential "muonness," while the other neutrino carried "antielectronness," to counterbalance the creation of an electron. In other words, rather than there being only one type of neutrino, perhaps there were two, an electron neutrino and a muon neutrino. It was a neat idea, but purely theoretical. The question was, how to test it?

The clincher came in 1968, when Steinberger teamed up with Leon Lederman and Melvin Schwartz and used Brookhaven's newly completed monster particle accelerator, the Alternating Gradient Synchrotron, to create a high-energy beam of particles called pions. The pion was known to decay into a muon and a neutrino. If Pontecorvo's theory was right, this neutrino should be a muon neutrino. After the pions had decayed, the resulting beam of muons and neutrinos was fired into a thirteen-meter-thick piece of steel made from an old battleship, which filtered out the muons but not the unstoppable neutrinos. Carrying on their merry way, the neutrinos then passed through a large aluminum target, whereupon a small handful of them collided with aluminum atoms and turned back into electrically charged particles.

Here's the punch line: in every case, these new particles were muons. In other words, the neutrinos really did carry "muonness." There were, it could safely be said, two types of neutrinos in the world, not just one. (These two neutrinos were later joined by a third, when an even heavier version of the electron known as the tau was discovered in 1976.[*])

The Kamiokande experiment in Japan could see both elec-

---

[*] By this point, the standard model of particle physics had taken shape, which required the existence of a corresponding neutrino for each of the electron, muon, and tau. Thus, the tau neutrino was born.

tron and muon neutrinos. The neutrinos from the Sun it detected were mostly of the electron variety. Muon neutrinos, meanwhile, came from a different source.

As we've seen, Earth is under constant bombardment from subatomic particles from outer space. Most of them were blasted into the cosmos by exploding stars, long, long ago. When they strike an atom in the upper atmosphere, they create a shower of particles, including a healthy dose of muon neutrinos. Not only could the Kamiokande observatory see these muon neutrinos, but it could tell them apart from their electron neutrino cousins. By the mid-1980s this led to yet another neutrino anomaly. Physicists had calculated that cosmic rays ought to produce about twice as many muon neutrinos as electron neutrinos, but measurements revealed they came in roughly equal quantities. Did this mean that muon neutrinos were *also* going missing? Or were there more electron neutrinos than expected? No one could be sure, but with two different anomalies on their hands, physicists increasingly began to wonder whether neutrinos themselves might be the cause of all the trouble.

To settle the matter, a supersized version of the Kamiokande observatory was built, and by 1996 it was ready to start taking data. Super-Kamiokande (SuperK for short) was ten times larger than the original, a humongous tank containing fifty thousand tons of pure water lined with 11,200 gleaming golden light detectors. What made SuperK particularly powerful was its ability to resolve which direction a neutrino had come from—a capability that allowed it to make a striking discovery. Not only were the muon neutrinos produced in the atmosphere disappearing; the problem was even worse when they had been created in the sky on the opposite side of the globe and flown *through* the Earth to arrive at the detector from below. In other words, the farther the neutrinos had traveled, the more of them seemed to get lost on the way.

If muon neutrinos were vanishing after only a few thousand kilometers, then electron neutrinos that had flown 150 million kilometers from the Sun could easily be disappearing en route as well. The only question was, well, where were they going?

One slightly bonkers explanation proffered was that they were slipping through extra dimensions of space and ending up in a parallel universe. A less outlandish explanation had been theorized by Bruno Pontecorvo as far back as 1957. But the problem was, his idea contradicted the tenets of the standard model of particle physics.

According to the standard model, neutrinos were massless particles that zipped through space at the speed of light. However, Pontecorvo realized that if neutrinos had even very tiny masses, then they should be able to do something no other fundamental particles could: change their identities as they traveled.

This effect is known as neutrino oscillation, and it's absolutely crucial to the anomalies we'll be discussing shortly, so it's worth taking a bit of time to understand. The basic idea is this: as we've seen, the neutrinos created in the Sun or by cosmic rays in the upper atmosphere come in three distinct types or "flavors," electron, muon, and tau, with the Sun predominantly creating electron neutrinos while cosmic rays mostly make muon neutrinos. However, according to Pontecorvo's idea, if neutrinos have mass, even just a tiny bit, then once a neutrino is created, it will begin to morph from its original flavor into a quantum mechanical mixture of all three. So, for instance, as an electron neutrino created in the Sun travels through space, it starts to evolve into a mixed state that includes a certain amount of muon and tau flavor. It's a bit like picking up a tub of vanilla ice cream at the store but getting home to find that it's spontaneously turned into a Neapolitan.[*]

How exactly does this bizarre phenomenon work? Well, it comes down to the fact that there are two different ways of describing neutrinos: in terms of either their flavors—electron, muon, tau—or what are called their mass states, labeled 1, 2, and 3. The flavors of the neutrinos, which are the states that they're created and destroyed in, are actually *mixtures* of the three mass states. And crucially, only the three mass states have a well-defined mass.

How can the same fundamental particles be described in different ways? Well, the situation is a little bit like how a movie projector works. Projectors only actually emit three basic colors of light—red, green, and blue—but by mixing these three colors in different ratios, the projector can create the illusion of more or less any color you like. A mixture of red and blue light makes our eyes perceive purple, even though the projector isn't actually emitting any purple light. Similarly, you can get the appearance of yellow by mixing red and green, or cyan by mixing green and blue. These mixed colors—purple, yellow, and cyan—are like the flavor states of the neutrinos, which are really mixtures of the fundamental mass states, akin to the red, green, and blue light emitted by the projector.

Now, since neutrinos are quantum mechanical particles, they travel as quantum waves that wobble up and down on a characteristic cycle as they zip through space. Crucially, the quantum waves of the three mass states wobble up and down at different rates, with the heavier mass states wobbling faster than the lighter ones. Let's say for the sake of argument that an

---

[*] For the ice cream nonexpert that's a mixture of vanilla, chocolate, and strawberry.

electron neutrino is 50 percent mass state 1, 40 percent mass state 2, and 10 percent mass state 3. Because the mass states wobble at different rates, if you happen to detect the neutrino at a distance where mass state 1 is close to a minimum on its cycle, while mass state 3 is close to its maximum, you would see a totally different composition, perhaps only 20 percent mass state 1, 60 percent mass state 2, and 20 percent mass state 3. In other words, it no longer looks like an electron neutrino. This is similar to how changing the relative amounts of red, green, and blue light in a projector beam would make your eyes perceive a different color. And if by chance, the new mixture lines up better with, say, a muon neutrino, then you'd be much more likely to see a muon neutrino in your detector than an electron neutrino.

The upshot of all this (admittedly difficult) physics is that by the time an electron neutrino emitted by the Sun reaches Earth, it actually has evolved into a quantum mixture of electron, muon, and tau. If your detector is only sensitive to electron neutrinos, or muon neutrinos, or tau neutrinos, some will inevitably appear to be missing.

The clinching evidence that neutrinos really do oscillate came from the Sudbury Neutrino Observatory (SNO)—an enormous geodesic sphere containing a thousand tons of heavy water, situated two kilometers underground in Ontario, Canada. Unlike SuperK, which was only sensitive to electron and muon neutrinos, SNO could detect the influence of all three flavors, allowing it to measure the total number of neutrinos reaching the detector from the Sun. When they released their first results in 2001, they likewise found a deficit of electron neutrinos. But with the addition of contributions from muon and tau neutrinos, the rate perfectly matched the predictions of the latest solar models.

At last it was clear: reports of the Sun's death had been greatly exaggerated. It was shining perfectly happily, thank you very much. It was the neutrino, not our local star, that was having an existential crisis.

The discovery of neutrino oscillations was truly momentous. The implication that neutrinos have mass forced physicists to revise the standard model for the first and only time since it was created in the early 1970s. What's more, it left many unanswered questions, among them, why are neutrinos so light? Although today we still haven't managed to measure the mass of neutrinos directly, we know that they are hundreds of thousands of times less massive than the next-lightest particle, the electron. Could it be that they get their mass in a different way from other particles? Even more exciting is the possibility that the three neutrinos we know and love have superheavy partners, particles that could potentially explain dark matter or even how matter got created during the big bang.

The discovery of neutrino oscillations heralded a new era of exploring the universe using nature's most perplexing building blocks. But even before they'd been conclusively observed, other results were already hinting that stranger discoveries might lie ahead.

Unexpected Guests

Deep within the sprawling Fermilab site, a side road runs through forest and shrubland to a small artificial hill. In one side of its grassy dome, a gray metal door marked "Caution—Radiation Controlled Area" opens into a chamber filled with racks of electronics and scattered with cardboard boxes of old computer kit. Though the space looks abandoned, the continuous roar of fans and pumps suggests that something is still going on beneath the hill. A trapdoor in the floor reveals what: a subterranean vault filled by a huge metal sphere, the three-story-high, eight-hundred-ton MiniBooNE detector.

Though now dormant, MiniBooNE* remains a star character in an unfolding neutrino drama, one that promises to upend our understanding of nature's most elusive building blocks. Inside the sphere are 250,000 gallons of ultrapure mineral oil surrounded by 1,280 amber orbs—light detectors—which once watched patiently for flashes of light in the dark liquid, a sign that a neutrino had made contact with an atom.

MiniBooNE sits on an invisible line that slices across the Fermilab site, starting at a target area about 550 meters to the south. Here, protons accelerated by the lab's fifteen-hundred-foot circumference "Booster" ring smash into a target with enough violence to break the mighty bonds holding their constituent quarks together, rupturing the protons into a spray of exotic particles. Among the subatomic shrapnel are the pions we discussed earlier, as well as kaons. Each is made from a quark and an antiquark glued together by the strong force. They are then magnetically funneled into a fifty-meter tunnel. As they fly through the empty space, they decay, transforming into muon neutrinos and their electrically charged muon cousins. Finally, a wall of steel and concrete absorbs everything except the neutrinos themselves, which whiz on unhindered, forming a beam that passes through steel, concrete, earth, and rock alike.

The purpose of MiniBooNE was to follow up on an anomaly found by a similar experiment, a thousand miles away at Los

---

* Which stands for Mini Booster Neutrino Experiment in case you were wondering.

Alamos, New Mexico, in the late 1990s. The Liquid Scintillator Neutrino Detector (LSND) had seen something that was hard to explain using our current understanding of the science: muon neutrinos transforming into electron neutrinos over far shorter distances than was thought possible. As we saw earlier, neutrinos can indeed change their stripes as they travel. But such effects are usually only apparent once they have covered relatively large distances. LSND was positioned a mere thirty meters from a source of muon neutrinos; they shouldn't have had enough time to perform their Jekyll and Hyde transformation by the time they arrived at the detector. Nevertheless, despite starting off with an almost pure beam of muon neutrinos, the team at Los Alamos saw strong evidence that electron neutrinos were mysteriously appearing in their detector.

Their result suggested something potentially earth-shattering: the hidden influence of yet another, *fourth* type of neutrino. We've already met the three standard flavors—electron, muon, and tau. Many theories predict the existence of an even more intangible variety, a so-called sterile neutrino. We previously talked about those in the context of explaining the two mysterious signals picked up by ANITA, the detector hovering over the South Pole. Could they be implicated in a second anomaly?

Neutrinos are almost completely cut off from our everyday world, interacting with ordinary matter only very feebly via the weak force. One of the most peculiar aspects of the weak force is that it pulls on particles more or less strongly depending on their "handedness." Now, particles obviously don't have hands, so what am I on about? Well, as we've seen, all matter particles have "spin," akin to a football spinning as it sails through the air. Depending on which *way* the particle is spinning compared with its flight direction, we refer to it as either left-handed or right-handed.

In the 1950s, physicists were shocked to discover that the weak force pulls on left-handed particles much more strongly than on right-handed ones. In fact, in the case of massless particles, it *only* pulls on left-handers, completely ignoring massless right-handed particles.

Thanks to neutrino oscillations, we know that neutrinos have mass, just extremely tiny ones. So far, at least, we have only ever observed left-handed neutrinos. If right-handed neutrinos exist, they would be entirely cut off not just from our observable universe but even from the weak force, and thus almost completely unable to exert any influence on ordinary matter.

These *truly* invisible neutrinos, as opposed to the *almost* invisible varieties we're familiar with, are therefore called sterile.

Okay, but why would the existence of sterile right-handed neutrinos be that big a deal? After all, every other matter particle comes in both left- and right-handed varieties. It would be kind of surprising if right-handed neutrinos *didn't* exist.

Yes, but that isn't the end of the story. As we've seen, the universe is dominated by dark matter, a substance for which the standard model can offer only an unhelpful shrug. Sterile neutrinos are a prime candidate to explain dark matter, particularly if they are much heavier than their ordinary neutrino cousins. Many theories also propose that sterile neutrinos may feel their own set of dark forces, allowing them to interact with other dark particles and perhaps act as a portal to this hidden world.

But if sterile neutrinos don't even interact via the weak force, then how would we know that they exist? Well, one way they could reveal themselves is by quantum mechanically mixing with the ordinary three neutrinos, influencing the way red, green, and blue light from our metaphorical movie projector mix together. This mixing could affect the rate at which the ordinary neutrinos oscillate into different flavors as they travel.

All of this is why the unexpected appearance of electron neutrinos in the LSND detector, when only muon neutrinos were expected, was potentially very exciting. It suggested the unseen presence of sterile neutrinos.

However, LSND's signal was only at about the 3.8-sigma level—in other words enough to be interesting, but not yet conclusive. To try to settle the matter, physicists set out to build a new experiment at Fermilab, based on very similar principles but with a higher-energy beam of muon neutrinos and a much larger detector. Thus, MiniBooNE was born.

To make the new experiment affordable (you'll sense a recurring theme here), the team behind MiniBooNE traveled to Los Alamos to scavenge some of the old experiment's light detectors, climbing down into its tank to retrieve around a thousand of the amber globes that lined its inner surface. Back at Fermilab, they were installed inside the spherical tank, later to be filled with mineral oil. Janet Conrad, one of the leaders of the experiment, recalled the incredible spectacle of the near-complete sphere lined with its constellation of gleaming light sensors. "We had these yoga mats where you could lay on the scaffolding and look upward," she said. "It was like a universe of tiny amber moons. Oh, it was so beautiful."

MiniBooNE began taking data in 2002, and within a few years it too began to see unexpected guests in its detector. Despite the experiment sitting in a beam of almost pure muon neutrinos, researchers were finding increasingly strong evidence of their electron neutrino cousins popping up. However, puzzlingly, the energies of these electron neutrinos didn't match what you'd expect if the LSND anomaly really was caused by the presence of a fourth flavor of neutrino. There were now two anomalies from two different experiments, but the simplest "sterile neutrino" explanation was starting to look less convincing.

What else could it be? Could it be that something even weirder and more exotic was causing the effect? Or perhaps— more likely, given the fortitude of the standard model—some other bit of misunderstood background that was contaminating the experiment?

In addition to presiding over Muon $g - 2$, Chris Polly had been involved in MiniBooNE. He is pretty convinced that their results remain solid, though he admits that limitations of the technology mean that there is a chance the anomalous signals MiniBooNE registered could have been caused by standard model particles. There are some signals that can spoof an electron neutrino. One particularly nasty example is where a muon neutrino smacks into an atomic nucleus, knocking it into an excited state, before the nucleus reflexively emits a high-energy photon (a bit like the way you might let out a yelp of surprise if you got hit in the face by a stray tennis ball). The photon released by the nucleus will then bash into another atom, converting into an electron-antielectron pair. If you're particularly unlucky, the electron and antielectron will end up traveling in the same direction and look, as far as the detector is concerned, like a single lone electron produced by an electron neutrino. That means you could easily mistake a muon neutrino exciting an ordinary atom with an electron neutrino appearing where it shouldn't.

So, despite firming up the case that something weird was

going on with neutrinos, MiniBooNE's results remained inconclusive. What was needed was yet another neutrino experiment.

Landing Four Planes at Once

It was becoming clear that to chart a path out of the confusion, the successor to MiniBooNE would need to be based on a completely different technology, one that would be able to tell genuine electron neutrinos apart from potential impostors. It was in 2007, just as MiniBooNE was revealing its first evidence for anomalous electron neutrinos, that a group of physicists began plotting the next step on the anomaly hunt. Leading them was Bonnie Fleming.

A professor at Yale, Fleming had spent many years working on MiniBooNE and was well aware of its limitations, in particular its inability to tell genuine electron neutrinos apart from photons emitted by overexcited nuclei. The problem, in essence, was that MiniBooNE was only able to see Cherenkov light (those visual sonic booms) produced when a particle broke the speed of light barrier. To tell if a signal had actually come from an excited nucleus, they needed a detector that could spot the nucleus as it (more slowly) recoiled from the blow imparted by the passing neutrino and the subsequent emission of the high-energy photon, like a cannon recoiling after firing a shell.

There was one such technology on the market that had the potential to revolutionize neutrino physics: a detector based on liquid argon.

Argon is a gas at room temperature but liquefies when chilled to extremely low temperatures, below −186 degrees Celsius. The plan was to fill a tank the size of a school bus with the super-cold fluid and stick it in the path of Fermilab's muon-neutrino beam. The research team's hope was that it would be able to discriminate between genuine electron neutrinos and potential confounding backgrounds.

The promise of liquid argon detectors lay in their potential to produce beautifully intricate 3-D images of the cascades of particles created by neutrino interactions. As an electrically charged particle plows through the liquid, it knocks electrons from their argon atoms, leaving a trail of positive argon ions* and negative electrons in its wake. Then a high-voltage electric field causes the trail of electrons to drift to one side of the tank, where they are picked up by a grid of crisscrossing conducting wires. Physicists can then tell where in the tank the electrons were produced based on where they hit the wire grid and how long it took them to drift across the tank. This allows the detector to reconstruct the entire path of a charged particle as it ionizes atoms in the liquid.

Crucially, this type of detector would allow physicists to see *everything* going on in the tank, including the presence of any slow-moving nuclei that wouldn't have been picked up by the earlier experiments. If Fleming and her team could pull it off, they might finally be able to determine whether the neutrino anomalies really were pointing to the existence of sterile neutrinos—or perhaps something even weirder.

However, it would be a long road to make the experiment a reality, and there were plenty of doubters in those early days. As Fleming told me, "In the beginning people were really skeptical of the technology. I felt like I was an entrepreneur."

---

* Ions are atoms that have either gained or lost electrons, leaving them negatively or positively charged overall.

Although some work on liquid argon detectors had been done in the United States in the 1980s, since then the only experiment using the technology was a European project called ICARUS, based at the Gran Sasso laboratory in Italy. However, ICARUS had proved a lot more time-consuming to set up than expected, which left many of Fleming's colleagues wary that the idea could be made to work.

Nevertheless, by 2009, Fleming and her deputy Bill Willis, a "senior statesman," as Fleming put it, who had done pioneering work on liquid argon detectors back in the 1970s, had persuaded Fermilab to green-light the project. Rather confusingly, their new detector would be christened MicroBooNE, thanks to its weighing in at a mere 170 tons (compared with the rather heavier 600-ton MiniBooNE).

The newly assembled MicroBooNE team now faced several daunting obstacles. First of all, they would need an incredibly pure supply of argon, because any contaminants would gobble up electrons as they drifted through the tank and thus spoil their measurements. The filters used to clean liquid argon degrade over time, and would have to be sent back to the manufacturer to be refurbished, at great expense. So the MicroBooNE team developed their own set of filters—Fleming compared these to the Brita water filters you might use at home—that could regenerate themselves. Not only did this save them a load of cash, it resulted in even higher purity than they'd planned for.

She recalled one particularly scary part of putting Micro-BooNE together: stringing the grid of wires that would detect the ionized electrons. "That was really nerve-racking, because if one wire breaks, it'll short out the whole detector, and if you're full of liquid argon when that happens, there's absolutely nothing you can do. So, we spent a lot of time trying to break wires and figure out what would happen if they broke."

After years of work, a big moment came in March 2013: the delivery of the giant cylindrical cryostat that would house the detector. This oversized thermos flask, which had been designed to insulate the super-cold liquid argon from the surrounding environment, had been manufactured in Wisconsin and was being transported by road to the Fermilab site. The day of the delivery, Fleming was sitting in the main cafeteria with Gina Rameika, MicroBooNE's project manager, speculating nervously on what route the cryostat might be taking. After a while the tension became unbearable, and Fleming persuaded Rameika to jump in the car with her and drive north out of the lab in search of the precious cargo. As they cruised the streets of suburban Chicago, they would occasionally spy a promising-looking vehicle, only to realize that it was a milk truck. Disappointed, they turned back toward Fermilab, but just outside the site they finally saw it: a caravan of vehicles surrounding a large flatbed trailer carrying the hulking cryostat. As the procession entered Fermilab, Fleming's colleagues lined the road while a postdoc played a bugle to welcome the heart of their experiment to its new home.

On August 6, 2015, the moment of truth arrived. The cryostat had been filled with thirty-four thousand gallons of liquid argon. It was finally time to power up. The detector was designed to work with a whopping 128,000 volts applied across the tank, necessary to make sure the ionized electrons drifted quickly toward the wire grid.

However, as they began to turn up the voltage, disaster struck.

"We were here in this room when we first turned on," Fleming told me when we spoke, gesturing around the control room at Fermilab's Wilson Hall. "But we couldn't get above 70,000 volts. We kept tripping off and we couldn't figure out why."

She and her colleagues knew that running at around half the design voltage would degrade the detector's performance, but they were terrified that if they pushed the voltage too high, they might break something. "High voltage is a bit like black magic," she said. "Part of it is being brave enough to ignore any wibbles you get on the high-voltage supply and keep pushing higher." After trying in vain to find the source of the problem, they eventually decided that they would have to keep running at half the design voltage, for the entire five-year run.

"It was a little disappointing for me," Fleming said, the frustration apparent in her voice. "I want to understand why!"

To make matters worse, it had become clear that about 10 percent of the wires used to detect the ionization electrons were unresponsive, suggesting that Fleming's nightmare of broken wires shorting out parts of the detector might have come true. The team managed to stick a camera through a porthole in the side of the cryostat and have a look around, but they couldn't locate the cause of the problem. With the tank filled with –186-degree liquid argon, there was precious little else they could do.

However, despite the lower-than-planned voltage and the dead wires, the team immediately saw that their detector was working. Beautiful, tendril-like tracks began to stream across their screens. "That was fan-tastic!" Fleming said, laughing. "I admit, when you stand in front of a review committee and they say, 'Is this really gonna work?' and you say, 'Absolutely,' on the inside you're like, 'I really hope this works!' So seeing that it did work was a big relief."

With the experiment running, now began the hard task of analyzing the data as it came in. One of the greatest challenges facing the MicroBooNE team was how to interpret the incredibly intricate 3-D images they were producing as charged particles cascaded through the detector. In the end, they opted for three parallel strategies, including the use of machine learning techniques to recognize patterns in their data (the same kind companies like Apple and Google use to tell you whether a photo on the internet contains a cat). Meanwhile, three teams were working to see if they could confirm the apparent electron neutrino signals that had been observed by LSND and Mini-BooNE. Yet another tested the idea that the anomalies were simply down to overexcited nuclei spitting out photons. Micro-BooNE's other co-spokesperson Sam Zeller described the process as like landing four different airplanes all at the same time.

But land them they did. On October 28, 2021, the Micro-BooNE team was ready to announce their long-awaited results. Would they reveal that LSND and MiniBooNE's anomalies were nothing more than photons from overexcited nuclei? Or would they confirm the earlier results, conclusively breaking the standard model of particle physics and ushering in a new era in our understanding of the universe?

Well, as it happened, neither.

When Fleming and her colleagues unblinded their results, after six long years of data taking and analysis, they found, well, nothing. There was no sign of a single electron's being produced by electron neutrinos. Nor did they see any evidence for photons coming from nuclei that could have fooled LSND and MiniBooNE.

What their results did do was inflict a serious wound on the idea that sterile neutrinos were behind the anomalies. If that were true, then they would have expected to see extra electron neutrino events, just as their predecessors had. If anything, though, they had seen *fewer* than expected.

If neither sterile neutrinos, nor the most obvious back-ground were the cause, then what exactly had LSND and Mini-BooNE seen?

The evening of the announcement, my Cambridge colleague and MicroBooNE physicist Melissa Uchida appeared on the BBC's flagship current affairs show, *Newsnight,* to discuss the significance of the experiment's findings. Talking to a rather bemused Kirsty Wark, the show's long-serving anchor, Melissa gamely explained that finding nothing at all was in fact tre-mendously exciting. The reason for her bullishness? Well, if neither sterile neutrinos nor excited nuclei were the cause of the anomaly, perhaps something even more revolutionary was.

In the Dark

"I mostly think we know nothing," Fleming said with a wry chuckle, from the vantage point of more than a year after MicroBooNE's unblinding. "You delve into any specific ques-tion, and you realize there's so much we don't understand. On cosmological scales we're just a tiny speck."

As I left the MicroBooNE control room to return to my rental car, it was hard to disagree with Fleming's remark. Despite all the progress we've made in science in the past few centuries, from unraveling the history of the cosmos to breaking matter apart into its most fundamental constituents, we really are still groping forward in the dark.

The realization that the world we can see and touch is only a tiny sliver of everything that exists really does put the achieve-ments of modern science in perspective. We've undoubtedly learned a huge amount about the visible world—the standard model with its twelve matter particles and three fundamental forces is one of humankind's greatest intellectual achievements. But still our understanding is limited to a mere 5 percent of the universe. The remaining 95 percent remains hidden, unknown, perhaps even unknowable. "Dark matter" and "dark energy" are mere labels that shroud our ignorance.

We're like the drunk looking for their car keys under a lamppost. Asked whether they're sure they lost their car keys here, they say, "No, I lost them over there," gesturing into the darkness, "but the light is better here." So many of our strategies for finding clues to the dark universe are lamppost searches; we look where the light is best, where the laws of physics make it possible for us to look, but with no guarantee of success.

Each new way of searching is like flicking on an extra lamppost. The more we can switch on, the more pools of light appear in the darkness, and the better our chances of finding answers. The neutrino anomalies are new and particularly bright lampposts, under which some physicists increasingly suspect the key to the dark universe may lie. But for all we know, we may still be those same misguided drunks.

As for the unflappable, evidently undetectable sterile neutrino, the biggest nail in its coffin came not from particle physics but from astronomy. In 2013, the European Space Agency (ESA) spacecraft Planck unveiled the most detailed image yet of the fireball that filled the universe until 380,000 years after the big bang. Analysis of its faded light allowed cosmologists to determine the properties of the early universe, including how many types of neutrinos were zipping about in the primordial fireball. The answer they got was a resounding three, one for each of the known neutrino flavors (electron, muon, and tau), apparently leaving little room for a fourth.

There is an altogether different signal that might explain the MiniBooNE and LSND anomaly: a so-called dark photon. We know that photons are particles of light, so the idea of a dark photon may sound contradictory, perhaps summoning images of the kind of sinister "unlight" conjured by evil sorcerers in many a fantasy novel. In reality, if indeed they are real, dark photons would be the messengers of a new force, one that allows dark matter particles to push and pull on each other. As we've seen, dark matter can't interact with ordinary matter very much at all, or we'd have already seen it. But there's nothing to say that the dark universe can't have its own set of secret forces.

My theoretical physicist colleague and resident neutrino specialist at CERN Joachim Kopp helpfully elaborated. One of the most promising remaining explanations for the anomaly, he said, is a sterile neutrino decaying back into an ordinary neutrino by giving off one of these dark photons. The dark photon then decays into an electron-antielectron pair, which MiniBooNE could easily have mistaken for a lone electron. If that's right, then a dedicated search for electron-antielectron pairs by MicroBooNE may be able to shed light (or should that be unlight?) on the matter. So, the coming years will be crucial.

That said, MicroBooNE is unlikely to be able to solve the neutrino riddle on its own. "The problem with MicroBooNE is that it's just a little bit too small," Kopp said. However, the experiment is actually part of a series of three new detectors that are gradually coming online. A few hundred meters to the south of MicroBooNE is a new detector known as the Short Baseline Near Detector (SBND).

And then there's the daddy of all liquid argon detectors, the aforementioned ICARUS. The brainchild of the Nobel Prize–winning Italian physicist and inventor Carlo Rubbia, it was built at the Gran Sasso lab and was the first large-scale experiment to demonstrate the power of liquid argon as a method of neutrino detection. After several years running, it has now been shipped across the Atlantic to be installed just a few meters away from MicroBooNE at Fermilab. Between them, SBND, MicroBooNE, and ICARUS will allow physicists to probe what's happening in their shared muon-neutrino beam at a range of different distances from the source, and together should finally allow them to resolve the twenty-year-old neutrino puzzle. But will it be a big breakthrough and "new physics," or will the standard model triumph once again?

I asked Kopp where he would put his money—new physics or the standard model?

"Of course I hope for new physics, but I'm also realistic enough to know that most anomalies go away. So maybe I'd put my money on the standard model because that way I'd be happy either way," he said, breaking into a laugh. Such caution is understandable—perhaps only more so because of the huge implications if the neutrino anomaly is confirmed. "People would very soon use whatever is discovered as a tool to make more discoveries."

The stakes are high, both for our understanding of nature and for the future of neutrino physics as a whole. Whatever happens, it's clear that nature's most elusive ingredients will continue to beguile, confuse, and fascinate for years to come. And if we keep searching, experimenting, and theorizing, there's a chance that one day, perhaps not too far from now, they will give us our first opening into the dark universe, setting us off on a thrilling new journey of discovery.

Chapter 7

Beauty and Truth

Physicists at the Large Hadron Collider have spotted fundamental particles behaving in ways that we can't yet explain. Are we seeing the first hints of a grander picture of the subatomic realm?

To a casual visitor, the commune of Ferney-Voltaire seems a pretty unremarkable place. Nestled among neat tree-lined fields near the French-Swiss border, its easy calm is only disturbed by a Saturday market and the intermittent roar of jet engines from Geneva airport. However, in the mid-eighteenth century, the sleepy village briefly became a place of radical thinking and controversial art thanks to the arrival of the French writer and philosopher Voltaire. After outraging Geneva's Calvinists with a scandalous poem satirizing the life of France's patron saint Joan of Arc, Voltaire moved to Ferney in 1759, buying a large estate and setting himself up as the local patriarch. His good works included building a church, establishing new cottage industries, and putting on theatrical performances, which were banned in the stuffy city down the road. Famous guests including Giacomo Casanova and Adam Smith came and went from Château de Voltaire, and Ferney's newfound buzz saw its population swell by more than a thousand. In gratitude, the village was renamed in Voltaire's honor.

Those days may be long gone, but on the edge of town, just around the corner from the E.Leclerc *hypermarché* and a drive-through McDonald's, a different kind of revolution may be under way. Drive past the large car park full of local residents loading their weekly shops and round the bend onto an unmarked road and you'll discover a cluster of gray industrial buildings, the only visible evidence of a huge scientific endeavor going on beneath Ferney-Voltaire's quiet streets. For a hundred meters below the surface is the largest machine ever built by humankind.

The Large Hadron Collider is a scientific instrument on an unparalleled scale: an enormous twenty-seven-kilometer ring that accelerates subatomic particles to within a whisker of the speed of light, before smashing them into each other inside huge, cathedral-sized particle detectors. These detectors record the subatomic shrapnel that comes flying out from the collision, with the aim of giving us a deeper understanding of the fundamental building blocks of the universe and the forces that bind them. Ferney-Voltaire is home to one of these giant detectors, the LHCb experiment—a twenty-five-meter-long, five-thousand-six-hundred-ton hulk made up of the most advanced detector technology in existence, capable of measuring the positions of thousands of subatomic particles to within a fraction of a millimeter as they zip through the detector forty million times every second. The experiment is run by an international team of more than fifteen hundred scientists, engineers, and technicians from across the globe, and since I started my career in physics as a PhD student in 2008, I've been lucky enough to count myself as one of them.

For the past few years, my colleagues and I at LHCb have been at the center of a scientific drama, one that maybe, just maybe could lead to a new era in physics. After a decade of determined but so-far-fruitless searching at the LHC for answers to some of science's biggest questions, we have been seeing hints that the particles produced inside the experiment are behaving in ways that defy our current best theory of particles and forces, the venerable standard model. The hope is that we are finally seeing cracks in this great theoretical edifice. Breaking the standard model would open a pathway to a new, grander theory and bring us closer to the ultimate goal of a complete description of nature at the fundamental level. The stakes, therefore, are huge, both for the scientists involved and for physics in general.

This story is deeply personal for me; much of my work in the past few years has been consumed by these emerging anomalies, not to mention that of dozens of my colleagues on LHCb and in the wider physics community. So, I am definitely not an impartial bystander to this drama, but I will do my best to lay it out for you in an unvarnished way. Through its ups and downs I hope to give you a sense of what it's like to work at the frontier of experimental physics, a place that's full of uncertainty, promise, and peril.

Voltaire once wrote, "Aime la vérité, mais pardonne à l'erreur." Love truth, but pardon error. As we'll see, those words couldn't be more apt for this still-unfolding tale.

Hints and Rumors

In the late spring of 2015, I got together with a few other members of the Cambridge LHCb group in a windowless meeting room in the Cavendish Laboratory at the edge of Cambridge to discuss what was rapidly becoming *the* hot topic in particle physics. After years of smashing particles together at the highest energies ever achieved, hints were starting to emerge of something unexpected. The evidence had been building slowly, starting with a measurement released by LHCb in 2013, that seemed to deviate from the predictions of the standard model. Then, in 2014, a second team at the experiment had reported a different but potentially related effect. The trigger for our meeting had come a few weeks earlier: a third result seemed to back up the 2013 anomaly. Individually, none of the measurements were precise enough to be able to claim that the experiment had really discovered something new, but taken together, a compelling picture appeared to be forming.

The question was, what should we, the Cambridge group, do about it? The anomalies had generated a surge of interest within LHCb, and scenting a major breakthrough, several university research groups were now moving to stake out the most promising research topics. While we were already working on some related measurements at Cambridge, we wanted to make sure we didn't miss the boat. After all, it's not every day that you get the chance to contribute to a discovery that could transform our understanding of fundamental physics. That's a prize that no physicist would want to pass up.

One of the more challenging aspects of being part of a large international project like LHCb is that your colleagues are both collaborators and competitors. You work together to build the experiment and maintain and operate it, often in close-knit teams where firm friendships form. But when it comes to analyzing the data with the ultimate goal of making discoveries, each research group will fight (usually not literally) to claim ownership of the most interesting topics. While there are a wealth of different measurements that can be performed with the data, not all measurements are equal. In truth, only a relatively small handful have serious potential to discover something new, and those areas tend to get very crowded. The trick that everyone is trying to pull off is to identify a promising measurement that no one has started working on, publicly and loudly declare your interest, and then quickly knock out some early studies to show that you're already several steps ahead of any potential competitors. Of course, that doesn't guarantee that no one else will muscle in on your area. During my PhD, I had found myself part collaborating, part competing with a team from Glasgow who was studying the same particle decay as me, leading to a tense debate over whose result would take precedence in the final paper (a debate that I ended up losing).

To help us figure out which measurements might shed the most light on the anomalies, we'd invited along a few colleagues from the Cavendish's theoretical physics group, some of whom were already working on theories to explain the results that LHCb had been getting. Among them was a young PhD student called Sophie Renner. As we sat around the conference table sipping teas and coffees, she laid out the current state of play from the experimental side, before moving onto the theories that she and her supervisor had been working on. Two of these included a fifth fundamental force—a pretty exciting prospect. After all, no one had discovered a new fundamental force for almost a century. More tantalizing still was an ambitious proposal to subsume the standard model in a grander theoretical framework that explained not just the anomalies but also why the fundamental particles we know about exist in the first place. The idea that the anomalies could lead to a more unified theory of physics was thrilling, although everyone in the room knew there was a lot of work ahead of us before we got to that point.

I had come to the meeting armed with my own proposals that I hoped might provide additional evidence for or against the anomalies. I'd spotted a few measurements we could make that should be sensitive to the same effects our colleagues had observed but that no one seemed to be working on yet. However, I wasn't sure how interesting they were from a theoretical perspective. During our discussion Sophie was able to help me pick out one measurement that looked particularly promising.

I left the meeting feeling more confident that my idea had legs, but with a lingering fear that someone might already be on the same track. Later that day, while digging through past presentations, I came upon some work from a colleague at Orsay in France that looked as if it were running along the same lines as what I'd proposed. Feeling disheartened, I sent off a polite email gently inquiring whether they were actually pursuing the study, expecting a firm "yes, stay off our turf." Instead, I got an amiable reply inviting me to chat about it over Skype.

We talked, and it turned out that while they had been considering making a similar measurement, no one was working on it actively, and they were happy to leave the field to me and the Cambridge group. We were in business.

But exactly what business was that?

To understand these anomalies, why they were starting to cause a stir in 2015, and what they might imply about the fundamental building blocks of nature, we first have to delve a little deeper into what the LHC and LHCb actually do. The first thing to understand about the LHC is that it's a classic one-trick pony: It smashes protons (or very occasionally lead ions) into each other, over and over and over again. That's all it does. This relentless subnuclear carnage happens forty million times every second, twenty-four hours a day, seven days a week for about nine months of the year.

You might then wonder how we can make lots of different measurements when we do only one thing. Well, this is where quantum mechanics comes into play. Smashing two protons together is an inherently unpredictable process, and there is a near infinity of possible outcomes. When two protons collide at the LHC, their enormous kinetic energy (the energy they carry due to the fact they are moving very fast) is converted into new particles. In other words, the LHC *makes* matter out of energy. This is why we accelerate particles in the first place; the closer they get to the speed of light, the more energy they have, and the more energy they have, the heavier the particles that we can potentially create.

However, it's impossible to say ahead of time which particles will get made in a given collision. Quantum mechanics only allows you to estimate the probabilities of one outcome or another, with some more or less probable than others. For instance, only around two in every billion collisions will create a Higgs boson, the final particle of the standard model that was discovered in 2012. That's why the machine produces so many collisions: it is only by rolling the quantum mechanical dice over and over that you have a decent chance of spying some rare and exciting particle buried in the data.

That said, while there are a large number of different particles you can search for and study at the LHC, at LHCb we're mostly interested in just one. The *b* in LHCb stands for "beauty," a name given to one of the six known quarks. The beauty quark (also known as the bottom quark*) is a superheavy version of the down quark that makes up all atoms, along with up quarks and electrons. Beauty quarks are fascinating little things; studying them is one of the best ways to search for the influence of

---

* When the b quark and its partner the top quark were first posited, there was an attempt to name them beauty and truth. However, the more prosaic bottom and top won out, which is what most physicists now call them. However, those of us who study b quarks would rather spend our time pondering beauty than bottoms.

hidden forces or new ingredients of nature. They are also produced in huge numbers by the collisions at the LHC, a fact that
the LHCb experiment was specifically designed to exploit.

What makes beauty quarks so fascinating? Well, it all comes
down to the fact they are unstable. The reason we don't find
beauty quarks inside ordinary atoms is that they live for only a
little over a trillionth of a second before decaying. And because
beauty quarks are rather heavy, with a mass around ten thousand times bigger than the electron's, they can decay into a wide
variety of other, lighter particles. There are literally hundreds of
different ways that beauty quarks can decay, and some of these
decays are particularly sensitive to the hidden influence of new
forces or particles. So, if we make precise measurements of the
properties of these decays and they come out different from
theoretical predictions made using the standard model, then
that can provide indirect evidence of the existence of something beyond our current understanding.

Beginning in 2013, this is exactly what my colleagues at
LHCb had started to see. The first anomaly to emerge related to
a specific type of decay where a beauty quark transforms into
a strange quark (that's its formal designation, not just a statement of opinion) along with two muons, those heavy cousins
of the electron. This process is extremely rare; only around one
in a million beauty quarks will decay in this way.

Now, it might be tempting to think of a particle decaying by
breaking apart into other particles. However, since fundamental particles like beauty quarks aren't made of anything smaller,
when they decay, they actually *transform* into new particles.
The new particles that come flying out didn't exist inside the
original particle; they are created as the original particle ceases
to exist.

If we think of particles as little marbles zipping around in space, that idea sounds rather hard to swallow, one thing simply transforming into another. But as we've seen, particles are not actually fundamental in a strict sense. Instead, particles are little quantum mechanical ripples in *quantum fields*, invisible, ever-present objects that fill every cubic centimeter of the universe. When you think of particles this way, the concept of a decay becomes easier to understand. What really happens when a beauty quark decays is the energy that's rippling about in the beauty quark field transfers into other quantum fields, which in the case we're discussing are the strange quark and the muon fields.

However, this kind of transfer of energy between quantum fields is almost never direct. A beauty quark can't just turn into a strange quark and two muons all by itself. Something else is needed. That something else is the weirdest of all the known fundamental forces: the weak force.

We've talked at some length about the weak force already. You'll recall that it's the only way in which neutrinos interact with ordinary matter. Well, the weak force also has a special ability that marks it out from the other known forces of nature: it can change one type of matter particle into another.

How does this happen? Just as the electromagnetic force is communicated by the electromagnetic field, which has an associated particle called the photon, the weak force is communicated by something called the W field, whose particle is known as the W boson. It's the W field that can pull off this unique morphing trick, making it particle physics' equivalent of the fabled philosopher's stone, once thought by alchemists to be able to transmute base metals into gold. Another way to think of the W field is as a bridge connecting different matter fields to one another. Energy can flow from one matter field, over the W bridge, and into another matter field, changing one type of matter particle into another in the process.

However, the W field isn't able to just turn any old particle into any other. There are specific rules that dictate which transformations are allowed and which aren't. For instance, the W field can only turn one type of quark into another type of quark. It can't turn a quark directly into an electron or a muon.

There are six quarks in total, which you can split into two groups. The up, charm, and top quarks form a threesome known as the up-type quarks. They all have a charge of $+\frac{2}{3}$. Then there are the down, strange, and beauty quarks, known collectively as down-type and having a charge of $-\frac{1}{3}$. The W field is able to turn any up-type quark into any down-type quark and vice versa. It can do this because the W field itself is electrically charged, coming in two versions with a charge of +1 and −1. Extending the bridge analogy, you can think of the positive and negative aspects of the W field as like the northbound and southbound carriageways, with quantum cars flowing along them in opposite directions.

To make this a bit more concrete, let's use an example. A common way for a beauty quark to decay is for it to transform into a charm quark, which you can represent by the diagram below:

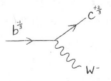

Here we start off with a beauty quark (b) that converts into a charm quark (c) by putting some energy into the W field. The important thing to notice is that the charge of the quark goes from $-\frac{1}{3}$ to $+\frac{2}{3}$, a total change of +1. It is a fundamental law of nature that electric charge cannot be created or destroyed, so to balance out the fact that the charge on the quark has increased by one, a negatively charged W boson (W⁻) has to be emitted.

It's for this reason that the W field cannot turn one down-type quark directly into another down-type quark. That would leave the electric charge on the quark unchanged, and because the W field carries electric charge, there's no way of keeping everything balanced in such a transformation.

But didn't I say that these anomalies relate to processes where a beauty quark turns into a strange quark? Both of these are down-type quarks, so how is this possible? Well, the answer is that you have to do something more complicated than emit a single W boson. You need a process that looks a bit like this:

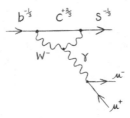

It may look like a mess, but what's essentially happening here is that the beauty quark emits a W boson (the wiggly line), turns into a charm quark, and then reabsorbs the W boson, which turns the charm quark into a strange quark. However, the beauty quark is much heavier than the strange quark, and that lost mass has to go somewhere. So, halfway along the W boson emits a photon (the second wiggly line labeled γ), which ultimately converts into two muons (μ). It is these two muons that carry off some of the difference in mass between the beauty quark and the strange quark.

If that gave you a headache, don't worry; the specific details of which particles do what isn't crucial. The important point is that turning a beauty quark into a strange quark is *compli-cated*, requiring a mixture of different quantum fields to work together to get you from A to B. If we think of the initial state (our beauty quark) and the final state (a strange quark and two muons) as two separate islands, then getting between them is like having to drive across several bridges connecting different intermediate islands before you reach your final destination. And as we saw earlier, the more complicated a process is, the less likely it is, quantum mechanically speaking, to happen.

One way to think about this is to imagine you have to pay a toll every time you use one of these quantum bridges. The more bridges involved in your journey, the higher the cost. However, since these are quantum bridges, the toll is paid not in dollars or pounds but in probability: the more complicated the journey, the less likely it is to take place at all.

The upshot is that a decay that involves a complicated mixture of fields is extremely unlikely. This is why for every million beauty quarks produced at the LHC, only around one will decay into a strange quark and two muons.

This infrequency might seem like a hindrance, but it is the very fact that these decays are so rare that makes them fantastic places to search for the hidden influence of new forces. Particle physicists have been searching for signs of new forces for decades, and one popular candidate comes in the form of a new quantum field known as the Z-prime. Like the W field, the Z-prime would be able to transform one type of particle into another, but unlike the W field it has no electric charge. This means it could turn a beauty quark *directly* into a strange quark, without needing help from any of the other fields. This is the equivalent of discovering a tunnel that gets you directly to your destination island, without any annoying stops at other islands. Now, the probability toll for using such a tunnel might be very high if the particle associated with the Z-prime field is very heavy. But as long as it's not *too* heavy, it could still compete with the complicated standard route.

The game we play experimentally is to measure the properties of these rare beauty quark decays as precisely as possible and then compare them with what theorists would predict using the standard model. The most obvious sign of a new force would be an unexpected change in how often the decay happens. However, more subtle effects are possible too. For instance, a new force could also alter the angles that the final particles come flying out at.

It was just such an effect that was found in the first anomaly paper back in 2013. My colleagues from LHCb made a careful study of the angles that the strange quark and the two muons emerged from the decay at and found noticeable shifts compared with what we'd expect based on our knowledge of the known forces. The paper that triggered our meeting in 2015 had made the same measurement a second time, but using almost three times as much data, and again, the same deviation in the angles showed up.

The 2014 result was something different, though, and potentially even more exciting. This time, my colleagues had compared how often a beauty quark decays into a strange quark and two muons with how often it decays into a strange quark and two *electrons*. Now, as we've seen, muons are carbon copies of electrons, with exactly the same properties except for the fact that they are around two hundred times heavier. As a result, all the forces in the standard model, including the weak force, pull on electrons and muons with the same strength. This principle is known as lepton universality (leptons are the category of particles that includes electrons and muons), and it's a hard rule of the standard model. This means that if you compare any two decays where you swap electrons for muons, they ought to happen at almost precisely the same rate. However, the 2014 analysis had revealed hints that the muon decay was happening only around 75 percent as often as the electron decay, violating the principle of lepton universality.

Big whoop, you might think. But for particle physicists, this was an astonishing result. There is absolutely no way to explain such an effect using the standard model. All three quantum forces treat electrons and muons the same—that's an breakable rule—and the only way you can account for a difference is if there is a new force we've never seen before that interacts with electrons and muons differently. And the reason that such a discovery would be truly earth-shattering is that it would most likely be linked to the reason that quarks, electrons, and muons exist in the first place.

One of the greatest mysteries still facing us is why the standard model is the way it is. We have six quarks—down, up, strange, charm, beauty, and top—and six leptons: the electron, muon, tau, and their three partner neutrinos. Then there are the three forces, each with its own associated particles that interact with some of these matter particles, but not others. We have no real idea *why* these particles exist; we simply observe them in nature and put them into the theory, a bit like a zoologist collecting species of butterflies.

However, there are tantalizing patterns, in particular the fact that the matter particles come in what physicists refer to as three generations. The first generation comprises the ordinary stuff that makes up the universe around us: the electron, the electron neutrino, the down quark, and the up quark. For reasons we can't yet fathom, there are two more of these so-called generations, which comprise heavier copies of the first-generation particles. The muon, for instance, is the second-generation version of the electron, while the beauty quark is the third-generation version of the down quark.

Why do particles exist in these repeating patterns? It's as if nature were telling us that there's some deeper principle at work here, perhaps some profound symmetry that connects all these apparently disparate fragments. But even though it might be staring us in the face, we haven't yet been able to discern from this structure a deeper truth.

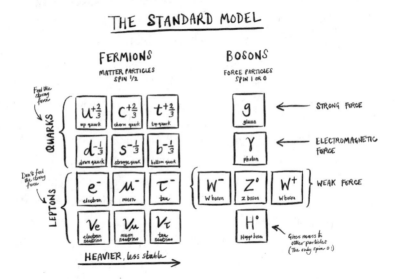

To come back around to that 2014 result: a force that breaks lepton universality and treats electrons and muons differently is thrilling not just because it violates our expectation but because it must somehow be connected to this deeper symmetry, part of a grander description of nature that explains why we have the basic building blocks in nature that we do.

And so this is why some of us were getting quietly excited in the summer of 2015. There was a genuine if remote prospect of finding the next layer of the cosmic onion.

However, as we know too well, all good scientists need a healthy dose of skepticism. A handful of hints is nowhere near enough to have any confidence that you're seeing something real. All the results so far were at or below the three-sigma level, meaning there was a decent chance that they were mere statistical flukes.

In the summer of 2017, however, a fourth result changed the game. A new measurement testing an independent but related set of beauty quark decays found exactly the same effect as the one discovered previously: muons were being produced less often than electrons.

One result could easily be a random blip, but two? The game was afoot.

The Force Awakens

On December 3, 2018, particles smashed into each other inside the LHC for the last time ahead of what would prove to be a three-year hiatus. In the CERN Control Centre, the engineers in charge of the huge machine safely disposed of the fearsome proton beams by activating a special set of kicker magnets, which knocked the two counter-rotating beams out of the ring down two six-hundred-meter tunnels and headlong into water-cooled targets made of hundreds of tons of carbon, steel, and concrete. As two loud bangs echoed along the empty tunnels, on the surface an image flashed up on a screen of the superheated arcs left by the beams as they scorched their way across the carbon targets, which heated to more than fifteen hundred degrees as they reeled from a blow with the power of 500 million stampeding horses. The second run of the LHC was officially over.

Later that evening, a winter thunderstorm knocked out power to large parts of the sprawling laboratory, and the engineers were forced to pause their planned maintenance on the superconducting magnets that keep the beams on their orbits. As Christmas approached, they began the slow process of warming the twenty-seven-kilometer ring up to room temperature from its operating temperature of –271 degrees Celsius, drawing out 130 metric tons of liquid helium that flows intravenously through the magnets and keeps them cool. Soon, people instead of particles were rushing around the ring, beginning an intense period of work designed to get the collider into tip-top shape for the planned third run in 2021.

Below Ferney-Voltaire, the first incarnation of the LHCb experiment had fulfilled its ten-year mission, providing us with a wealth of new information on the behavior of the building blocks of matter. Our collaboration now had a daunting task ahead of it: to dismantle the huge instrument, hauling tons of material up the hundred-meter access shaft to be put into surface storage, while a race began belowground to install an almost entirely *new* version of the detector before the LHC began colliding again. *LHCb II: The Bottom Strikes Back*, as it would have been known in a better world (instead of being rather dully referred to as the LHCb Upgrade), would be far more than an upgrade. It would be a brand-new detector, with most subsystems replaced entirely and the remainder getting a significant revamp. All of this work had to be squeezed into just two years, an incredibly tight schedule made all the more pressured by the knowledge that the greater LHC would not wait if our part of it wasn't ready when the time came.

Work was also being done on the other three LHC experiments, but only LHCb was going through a total overhaul. However, the payoff for all this work and stress would be an experiment that could record data at five times the previous rate, allowing us to tease out ever rarer and more exotic processes while testing the standard model to the breaking point. For those of us working on the anomalies, most exciting of all was the prospect of finally having enough data to rule our findings in or out once and for all.

However, progress on the anomalies would not have to wait for the upgrade to start taking data in two years' time. The last collisions of 2018 had now flowed through the surface computer center and out across the Worldwide LHC Computing Grid—a global distributed network of 1.4 million computers that stores and analyzes the data produced by the LHC experiments. A vast ocean of pristine data sat waiting to be analyzed. In its depths lay the promise of discovery.

There were multiple teams itching to get their hands on the recently recorded data and make more precise measurements of anomalies, in particular to test the principle of lepton universality—that forces should treat electrons and muons the same. The teams had already spent several years preparing studies of how often beauty quarks decay into a strange quark and either two muons *or* two electrons. At the end of all the painstaking data analysis, they would arrive at a single number: the ratio of the number of muon-producing decays divided by the number of electron decays. According to the standard model, this ratio, referred to as R, should be almost precisely equal to 1, which is another way of saying that the number of muon decays should be equal to the number of electron decays. Any deviation from 1 would be unambiguous evidence for a new force of nature.

Now, because all the teams were trying to measure the same thing—this R ratio—you might ask how they could all make independent measurements, even as they drew from the same data. It all comes down to the fact that you never see a quark on its own; they are always bound up tightly with other quarks to form composite particles, which are what you actually see in the detector. When a beauty quark is produced in a collision, it can grab onto any of the four lighter antiquarks (up, down, strange, charm) to form a quark-antiquark pair, or alternatively get together with two other quarks to form a threesome akin to a proton.* Adding to this diversity, the strange quark also ends up in a composite particle after the decay, and this can be in a high-energy "excited" state or a low-energy "unexcited" one.

All this ultimately means that while there's only one fundamental decay going on deep down, there are numerous different signatures you see in the experiment, depending on what other quarks the beauty quark is stuck together with. It's a bit like having a single type of car engine shoved into multiple different bodies. On the outside the cars look different, but under the hood they're the same machine. This means that each team could look at a different set of composite particles, with each of them offering an independent way to measure the same fundamental process.

First out of the blocks was a team led from Imperial College London. Having pioneered the work on this set of beauty quark decays, they were now working to update their original 2014 measurement with additional data recorded in 2015 and 2016. They are known as the RK team after the ratio that they were trying to measure, where the K bit refers to the type of strange-quark-containing particle that gets spat out by the decay.

In the meantime, a rival team led from the Centre National de la Recherche Scientifique in Paris, who had been involved in the 2017 measurement of a ratio called R_{K^*} (spoken as "R-K-

---

* A proton is made of three quarks—two up quarks and a down quark.

star," where the asterisk denotes that the strange particle is in an excited state), was gearing up to make their own updated measurement. Their plan was to use all the available data recorded through the end of 2018.

What raised the hackles of the RK team is that the Parisian team wanted to measure both the R_{K^*} and the R_K ratios *at the same time*. Their argument was that measuring the two different ratios at once would make it easier to deal with shared potential biases, leading to more reliable final results, an argument that the RK team took issue with. This second group became known as the RX team.

Safely out of the line of fire, I was deep into work at Cambridge (by myself, for the time being) on two other ratios, called R_{K_s} and R_{K^*+}. Neither had been measured at LHCb before, but in principle they should be sensitive to the same effects if a new force was really out there. I had managed to carve out this area for Cambridge, but that's in part because measuring these second two ratios is much more challenging than measuring the others, because the strange particles produced in the two decays I was studying have the unpleasant habit of flying out of the detector unseen. I knew from the outset that my measurements wouldn't be able to compete statistically with the main R_K and R_{K^*} results, but I hoped that they could shed a bit more light on the mystery and perhaps turn up hidden biases, if there were any.

However, I soon realized that I had bitten off more than I'd reckoned with. Testing lepton universality at LHCb is maddeningly complicated thanks to the simple fact that the LHCb detector is much better at seeing muons than electrons. Muons are dead easy to spot. Thanks to their large masses, they pierce the detector like a high-velocity bullet, leaving a telltale electrical signal at the outermost edge where no other particles can reach. If you see a signal at the edge of the detector, you can be pretty darn sure it's a muon.

Electrons, on the other hand, are a pain in the arse. These irritating little lightweights get knocked all over the shop as they scatter off atoms in the detector, not so much bullets as tiny subatomic pinballs, causing them to radiate away some of their energy before they get absorbed in the inner part of the detector. Not only does this mean you generally mismeasure their energies, but it also makes them easier to mistake for other particles. Unlike the edge of the detector, the inner detector is a busy place, constantly bombarded by particles made of quarks, which can make telling electrons apart from all this quarky background tricky.

The upshot is that the detector is about four times better at spotting muons than electrons. Clearly this is a big problem if you're trying to compare the number of muon decays with the number of electron decays; if you don't properly account for the fact that the detector is better at seeing muons than electrons, you'll get a ratio that is way off the expected number—1—and will conclude that you've made a Nobel-worthy discovery when in fact you've just measured the degree to which your detector is biased.

To make these sorts of measurements at all, you have to overcome two big challenges. First, you have to precisely determine how good the detector is at spotting electron and muon decays. And second, you need to estimate how often you might mistake other types of particles for electrons or muons. Take both of these into account, and you can make an unbiased measurement of the true ratio R. In practice, this is accomplished using a combination of real data alongside large quantities of simulated data. These simulations are incredibly impressive, re-creating a virtual model of the entire detector down to each individual sensor (of which there are millions), shooting simulated particles through it and producing data that looks as close as possible to the real thing. And because it's a simulation, you know exactly what particles you started off with, so you can use it to determine how the detector performs.

The quantities of data and simulation required to make these measurements are daunting enough; just producing the files you need can take months, even using the immense computing resources afforded by the LHC Worldwide Computing Grid. And that's before you even get into the real meat of the analysis. Both the RK and the RX teams numbered around ten or more members.

Fortunately, in late 2017, I was joined by a newly arrived PhD student, John Smeaton. Not only was John enthusiastic and good-humored, but he was also intimidatingly able and was soon putting out the kind of work I'd expect from someone years further into their career. I realized that together we had a decent shot of being able to pull the whole thing off.

However, neither of us realized back then how long the journey would be.

After the flurry of interest in lepton universality following the studies in 2014 and 2017, the first moment of truth would come toward the end of 2018. The RK team based out of Imperial College London had completed their analysis of the data recorded in 2015 and 2016, and—after a long period of scrutiny by the collaboration and a specially appointed review committee, who drilled deep into their methods—they'd finally been given permission to begin the process of unblinding.

They had spent the past few years improving almost every aspect of the measurement, introducing new cross-checks and ways to correct biases caused by the detector. Paula Álvarez Cartelle, who was a postdoc at the time and shared the bulk of the work with a PhD student, told me that at the outset she was pretty convinced the anomaly would disappear. "The idea was to improve the precision and finally get rid of this lepton non-universality nonsense, basically to correct the mistake and move on," she said.

But as she got into the "gritty bits" of the analysis, the team realized that the measurement was, in her words, "super robust." In particular they put a lot of effort into understanding LHCb's differing ability to detect electrons and muons, which would have been the obvious explanation for the observed anomaly. But after years of searching for a mistake, they couldn't find one.

The big question was, would the new ratio agree with the value that had been found in 2014, or would it move back toward 1, the prediction of the standard model?

The first step in the unblinding process was to reanalyze the data that had been used in the 2014 result and check that they arrived at the same answer—to essentially replicate the earlier work and ensure that it had been done accurately. If they did, then they'd get the green light to unblind the new data as well. If not, it would suggest a mistake in one of the two analyses, introducing a long delay as they were forced to pick over the potential cause.

To the team's relief, when they unblinded, it came out almost exactly the same: $R = 0.72$, with an uncertainty of about 0.08. The earlier measurement held up.

On its own, the result was already over three sigma from the standard model. Suddenly the stakes of their work skyrocketed. Paula and the team realized that if the new data gave a similar value, they'd cross the critical five-sigma threshold, meaning they'd have discovered the influence of a new force of nature.

"We were like, shit, we are going to get like five sigma here! Oh my God, we have it," she recalled.

It was an anxious overnight wait before they got permission to unblind the final stage. The next day, just after noon on December 7, they got back together in their office at CERN and pressed the button. Everyone's faces fell when the result appeared on the screen: 0.93 with an uncertainty of 0.09. The new data had shifted decisively toward the standard model prediction of 1, agreeing within just one sigma.

Dreams of a major discovery evaporated in an instant. It would be a few months before the result was made public, but anyone who saw the RK team in the CERN cafeteria that day could read the result from their faces. Mitesh Patel, who led the RK team, recalled, "We just felt emotionally train wrecked. We were gutted."

When they combined the new data with the old data, they still got a value of 0.85, with an uncertainty of 0.06, tantalizingly *just* shy of three sigma. What's more, there was further data to add, which might swing the result back the other way. It wasn't game over just yet. Mitesh told me that after the initial disappointment "we picked ourselves up, dusted ourselves off, and got a team together to add the 2017 and 2018 data."

Back in the U.K., I had seen the result come through via email as I was leaving a meeting in central London. I walked past the British Museum squinting at my phone in the bright winter sunshine, trying to make sense of what the RK team had found. One thing that became clear was that our own measurements in Cambridge had gained new importance. An inconclusive result like this meant that any extra evidence we could provide might tip the balance. I also felt a sense of relief: had the anomaly disappeared altogether, then years of work would have been wasted.

When the result became public a few weeks later, most informed observers expected the lepton universality anomaly to disappear in time as further data was assessed. Not everyone was so pessimistic, though.

Gino Isidori, a theoretical physicist at the University of Zurich, had spent the past few years building theories that could account for the anomalies. For him, the original lepton universality anomalies had been quite hard to accommodate because they were *too big*. However, a value of around 0.85 was actually much easier to explain using new particles or forces, without breaking the theory's agreement with data from other experiments. "I was not too disappointed," he told me. "In many ways I couldn't have wished for a better result!"

The reason for Isidori's optimism was that he was thinking about more than just the anomalies in beauty decays to strange quarks. Hints of the breaking of lepton universality had also been emerging in a separate class of decays, where a beauty quark turns into a charm quark and emits either a tau or a muon (the tau is an even heavier copy of the muon). Years earlier in 2013, BaBar, an international experiment based at Stanford, had seen evidence that the decay to taus was happening more often compared with muons than you'd expect in the standard model. Their results soon received support from measurements by the Belle experiment in Japan, as well as from LHCb.

Rather than just one or two anomalies, a whole host seemed to be emerging from related areas of particle physics. It was as if we were spying the footprints of a great, unknown beast as it crashed about in a dark jungle. What excited Isidori and his colleagues was that they had discovered it was possible to explain all of these different anomalies *at the same time* with a single theory. And more exciting still, this theory promised to revive the long-stalled grand mission of physics: unification.

As we've already seen, there's a lot about the standard model

we don't understand. Why are there six quarks and six leptons? Why do the matter particles come in three increasingly
heavy generations? And why are there three different forces of
nature? Since the standard model was first assembled in the
1970s, physicists have been searching for a more unified theory
that would bring the forces together and explain the origins of
the matter particles. There have been several great unifications
in the history of physics. Isaac Newton unified the earthly force
that pulls apples from trees with the celestial one that holds
the Moon on its orbit. In the nineteenth century, James Clerk
Maxwell showed that electricity and magnetism were different aspects of a single unified electromagnetic force. Finally,
in the 1970s and 1980s, physicists realized that the weak force
and electromagnetism were part of a more unified electroweak
force, giving birth to the standard model in the process. Buoyed
by their triumph, theorists soon attempted to unify the strong
force, the weak force, and electromagnetism into a single force
(which in my view ought to be called simply the Force). They
were seeking what became known as grand unified theories. At
the heart of these theories were beautiful symmetry principles
that transformed the ad hoc standard model into a fragment of
a larger, more elegant description of nature.

However, by the late 1980s the unification project had hit
the buffers. The predictions of grand unified theories failed to
be verified experimentally.* Another problem was that all of
the new particles predicted by these theories had enormous
masses, far beyond the reach of the world's most powerful
particle colliders, making them all but unknowable. While the

---

* A major blow came from the Kamiokande experiment in Japan that
had failed to find evidence for protons decaying, a key prediction of
grand unified theories.

yearning for unification remains deep in the hearts of many physicists, the community began to believe that such lofty goals would not be achieved in the foreseeable future, and the quest for unification fizzled.

However, the anomalies in beauty quark decay had spurred Isidori and others to look at these ideas anew. To their amazement, they found that some of the concepts from grand unified theories of the 1980s could be repurposed in a way that no one had considered at the time. Back then, theorists had been focused on the unification of the forces, but if instead you focused on the unification of matter particles, like leptons and quarks, a new and compelling picture emerged.

According to these new theories, the standard model could be a low-energy echo of a far more symmetrical theory that lives at very high energies, like fragments of patterned glass fallen from a high stained-glass window. This profound symmetry is hidden from us at the energies we've so far been able to explore, but go high enough and you will discover a more perfect order to the ingredients of nature.

What distinguishes quarks from leptons is that quarks feel the pull of strong force, while leptons (electrons, muons, taus, and their partner neutrinos) do not. The property that allows a quark to feel the strong force is called color, which is the strong force equivalent of electric charge. However, while there is only one type of electric charge, color comes in three varieties known as red, green, and blue (although "color" is just a label; it has nothing to do with actual everyday color).

In the standard model, leptons don't carry color and so don't feel the strong force. But what if instead of three colors, there were four? This extra color—often called lilac—could be carried by the leptons, effectively making them into quarks and thus unifying the matter particles. Better still, the strong force could be a fragment of a more fundamental force that pulls on all four colors. However, at everyday energies, this grander force breaks apart, leaving us with the three-color strong force we know and love, along with a leftover single-color force. And a force with a single color looks just like the electromagnetic force with a single type of electric charge, meaning this picture could also help unify the strong force with the electromagnetic and weak forces.

This was seriously heady stuff. If Isidori and his colleagues were right, then the beauty anomalies were about to usher in a revolution in particle physics of a type not seen for decades. Unification was back on the cards, and with it the once-distant goal of a final theory of nature came a significant step closer.

However, before we got ahead of ourselves, we had to remember that none of these anomalies was convincing on its own. The threat of hidden biases, experimental or theoretical, loomed large. The great ship of progress could yet hit the rocks of human error or cruel statistics. We needed more data.

A New Dawn

At the start of 2020, as the COVID-19 pandemic sent the world into lockdown, work on the LHCb upgrade slowed to a crawl. Restrictions on movement made transporting new detector parts to CERN from universities across Europe a huge logistical headache, and the potential harm of sharing confined spaces seriously disrupted work in the cavern. But for those of us analyzing LHCb's existing data, things continued more or less as normal, only with more Zoom meetings and less work-appropriate attire.

Nor did COVID stop the production line of new physics

results. In March 2020, LHCb released a long-awaited update of a measurement of the angles that the strange quark and muons come flying out at when a beauty quark decays. The previous result in 2015 had shown a 3.4-sigma tension with the standard model. However, despite what on the face of it looked like convincing evidence, many were skeptical that the standard model predictions for these angles could be trusted.

With any anomaly where an experimental measurement disagrees with a theory prediction, there are broadly four possible explanations:

1. The result is a statistical fluke.
2. The experimenters made a mistake.
3. The theorists got their sums wrong.
4. You've discovered something! Have a biscuit.

It's only when you eliminate the first three boring explanations that you can be sure you've made a genuine discovery. Physicists try to mitigate the first explanation—a statistical fluke—by simply accumulating enough data and setting a stringent test for what you call a discovery. This is the job of the five-sigma threshold. The next two—an experimental balls-up or a theorist forgetting to carry the 1—are often trickier to eliminate. It's one of the reasons lepton universality anomalies were so exciting: They are incredibly theoretically clean. What do I mean by that? As we saw with the muon $g - 2$ anomaly, calculating a prediction from the standard model is often extremely challenging. However, by taking a ratio of two similar decays that differ only in swapping muons for electrons, all the theoretical nasties cancel out and leave you with a pristine prediction—in this case, simply $R = 1$.

However, predictions of the angles at which particles emerge from the decay are more vulnerable to unknown effects from quarks and gluons. In fact, since the last result in 2015, theorists had worked hard to improve their understanding of the decay and, in so doing, had caused the original anomaly to weaken from 3.4 sigma to 2.4 sigma, demoting it from "evidence" to more of a suspicious whiff.

As a consequence, the angular anomalies being presented in March 2020 were generally given less weight by the community, even though they were still intriguing.

It was my colleague Eluned Smith who shared the new result before a sparsely populated council chamber at CERN as many more watched online. Adding the new data from 2015 and 2016, the angular anomaly had grown to 2.7 sigma, inching back toward the 3-sigma "evidence" barrier. For anomaly chasers, this was yet more grist to their mill. Others remained skeptical about the promise of a discovery. The veteran CERN theorist John Ellis told me, "I have a wait-and-see attitude; it looks to me like these anomalies might be going away . . . or at least they're not strengthening as one might hope."

Meanwhile at Cambridge, crunch time was approaching for John Smeaton and me. We had hoped to get our measurement over the line in 2020, but we had run into a couple of roadblocks that proved difficult to overcome. One was a cross-check of the measurement, where we took a ratio of *other* particle decays that involve electrons and muons—decays where we already knew the ratio was definitely equal to 1. If we applied our analysis to these decays and they returned 1, then that would give us some confidence that our methods were on the right track.

That's not what we found, though. Our ratio kept coming out at around 1.4. For months, we wrestled with the problem.

It was only by rebuilding a major part of the computer code we had built from scratch that we found the source of the issue and our cross-check returned measurements of 1.

Along the way, however, another problem arose, one that foreshadowed events to come. John had found that there was an unexpectedly high chance of mistaking particles known as pions for electrons. As I mentioned, electrons can occasionally be confused for other particles made from quarks. And since we were comparing electrons with muons, anything that could masquerade as an electron was extremely dangerous. John had found that we'd expect around 10–15 percent of the electrons we recorded to actually be pions in disguise. It was a troubling figure, since the latest value of R measured in 2019 was only around 15 percent below 1. Could the whole anomaly story we'd been chasing really just be down to us confusing pions for electrons?

After a lot of detailed work and some back-and-forth with the RK team at Imperial College London, we convinced ourselves that the pions really *were* under control. RK had taken them into account in their analysis and were sure they were negligible. Further studies of our own showed that the rate of fake electrons was low enough to handle.

By the end of 2020, after three years of work, we were finally getting close. But the RK team had one final result to reveal first.

On January 25, 2021, with the U.K. back in yet another lockdown, the RK team assembled on Zoom to unblind their final result, including all the available data recorded by LHCb to date. "I was literally shaking," Mitesh, the head of that team, told me, as he waited for Paula to run the code and reveal the final result. The faces of his nervous colleagues looked back at him from his computer screen as he sat in his home office in

Putney, Southwest London. The stakes now were enormous. If the value of the ratio stayed still, the shrinking error would push them over the three-sigma threshold, back into evidence territory.

"You don't know what you're going to be dealt," he said. "You just do the best measurement you can. Opening the box is like looking at the face of nature for the first time. We didn't know what she had to tell us."

Paula, who had recently joined us at Cambridge as a newly tenured lecturer,* sat alone in her new apartment running the final code. When she read out the result to the eager team, there was, as Mitesh put it, an awful lot of swearing. The value of R_K hadn't moved a muscle. It was still 0.85. But the added data meant that the uncertainty had shrunk to just 0.4, crossing the three-sigma threshold.

They had discovered evidence for a new force.

It was a moment that every physicist lives for, the moment you learn something new about the world that no one knew before. When you open the box, as Mitesh put it, the world bifurcates: into one where you've made a breakthrough and one where you haven't. For him, it was the length of the journey that gave that moment its significance: "All those struggles, all the arguing with the review committee, last minute panics—all that adds poignancy to the moment."

Paula's reaction was just as joyful, if tinged with anxiety. "We felt the weight of the responsibility. At that moment you see the result and you think, 'This could have very large implications.' You don't want to think, 'I just broke the standard model,' but at the same time you're a bit, 'Oh shit!' "

* A lecturer is the equivalent of an assistant professor in the United States.

On March 21, 2021, Dan Moise, the PhD student at Imperial who had done much of the analysis work, revealed the new result to the world at the prestigious Moriond conference, high in the Italian Alps. Meanwhile, his colleague Kostas Petridis made a simultaneous presentation to a rapt CERN auditorium. The result caused a media sensation. Mitesh found himself on the BBC's flagship *Today* program explaining the result to millions of listeners, and he, Paula, and Kostas spent much of the day fielding interviews from journalists and TV crews. Mitesh recalled a frantic cycle across London to record a BBC TV interview: "It was the middle of lockdown and I hadn't cycled for months and months and months, but I got on the bike, cycled forty-seven minutes from home to Regent's Park, and arrived absolutely pouring with sweat, totally breathless, to record this interview."

While the team was careful to add the caveat that it was not yet a true "discovery" and there was always the chance of a missed error, it was hard not to get swept along on the wave of excitement. "It was extraordinary. Nothing like that's ever happened to me before and probably never will again," Mitesh told me. After a decade of null results at the LHC, many of us dared to dream that a new dawn was finally breaking.

The prevailing mood of optimism only grew when, just three weeks later, the Muon $g-2$ experiment in the United States announced their unrelated measurement of the muon's magnetism, confirming yet *another* long-standing anomaly and strengthening the sense that a revolution really was under way.

John and I would soon face our own moment of truth. On September 8, 2021, after getting the final green light from the team of physicists assigned to scrutinize our work, the two of us connected on Zoom to unblind our ratios—mercifully without the glare (yet) of onlookers. It was the end of a long journey, one that had begun more than six years earlier in a stuffy meeting room in Cambridge and that had involved countless hours of painstaking work. The fact that we had got to this point was testament in particular to John's dedication and skill. He had carried much of the analysis as I was spread among multiple projects. All that remained now was to open the box and see what lay in store.

I had known from the start that our results could never have as big an impact as the more statistically powerful ratios measured by the RK and RX teams, but when I saw the numbers, I couldn't help but feel a thrill. Both of our ratios had come out at around 0.7 with an uncertainty of about 0.2. They were only at the 1.5-sigma level, but crucially, they lined up perfectly with the previous results; in other words, they too supported a break from lepton universality.

After taking a moment to congratulate each other and marvel at our two hard-won numbers, we fired off an email to LHCb management. Niels Tuning, who was in charge of coordinating the LHCb physics program as a whole, came back to us quickly: "I have goosebumps!"

A few weeks later, I flew to CERN for the first time since the pandemic had begun to present our results to the outside world. I was to speak in the same auditorium where the discovery of the Higgs boson had been announced ten years earlier. I was under no illusion that our results had the same import, but still I felt as if I were part of something big. As I took the stage in Switzerland, John was giving a simultaneous talk at a conference in the beautiful Château de Blois in the French Loire valley.

At the workshop at CERN that followed my presentation, it became clear that ATLAS and CMS—the two biggest experiments at the LHC—were starting to pay attention to what we were up to and thinking hard about how they might search directly for new particles that could be the source of anomalies. A big moment came when Fabiola Gianotti, the director general of CERN, connected virtually to hear the talks.

But perhaps the most significant outcome of the workshop was that multiple theory groups now agreed that while no individual result counted as a discovery, when combined they created a deviation that was well over five sigma. Could the standard model be cracking before our eyes?

At the end of that October day, I finally had the chance to share a beer with some of my LHCb collaborators in the CERN cafeteria. It was only then that I realized the degree to which our result really had moved the dial. One colleague told me that when he saw our two numbers, he had instantly thought, "Nature is speaking to us!" I flew back to London with a sense that I really had been part of something that might change how we see the world.

But the euphoria was not to last. As I stood on that stage at CERN, the RX team in Paris was getting closer to the moment when they would unblind *their* measurements, which combined the two most powerful ratios. However, first they would have to compare their R_K result with the value found by Mitesh, Paula, and their team earlier that year. It was there that the trouble began.

The Unraveling

During a lecture at Cornell in 1964, Richard Feynman gave a characteristically pithy summary of the scientific method: "If it [a theory] disagrees with experiment, it's wrong." To growing laughter from the audience of rapt students, he continued in his exaggerated Queens accent. "It doesn't make any difference how beautiful your guess is, it does not make any difference how smart you are, who made the guess, or what his name is—if it disagrees with experiment, it is wrong."

Now, who am I to argue with Feynman, one of the greatest minds of the twentieth century. But there is a possibility he forgot to mention: experiment can be wrong too.

Experimenters are by temperament a cautious bunch. For an experimental scientist there is no greater terror than realizing that a result you've put out into the world contains an error. This is all the more true when your result challenges the most successful scientific theory in history and ignites the imagination of the wider scientific community. Beyond the overwhelming desire to get it right, careers, reputations, and entire scientific fields can be on the line. So, while it may seem nuts that getting a single number from some data can take years of several people's careers, that's the forensic level of care and scrutiny you absolutely have to put in if you really want to find out something new about the universe. The hope is that all the long years of painstaking analysis, checks, and criticism will produce an experimental measurement that is solid and can be trusted—that reflects how the world *is*. But even after all that, things can still get missed.

The first clue that something might be amiss with the lepton universality anomalies came as winter gave way to spring 2022 and the RX team began the long-awaited process of unblinding their results. They had spent the past five years performing a combined analysis of two ratios, each of which was designed to test whether beauty quarks decay into electrons and muons at the same rate, as the standard model predicts. One of these ratios, R_K, was what had been measured just a year earlier by the RK team led by Mitesh and Paula. It had revealed a three-sigma tension with the standard model and triggered that attendant surge of excitement in the world's media. The other ratio, R_{K^*}, had been measured only once before, back in 2017, and using only a fraction of the now available data.

Leading the RX team was Vladimir Gligorov, a physicist based at the Centre National de la Recherche Scientifique in Paris. Known as Vava to his LHCb collaborators, he first got involved with the anomaly saga after chairing the committee that reviewed the R_{K^*} 2017 measurement, before it was published. Like R_K, that earlier measurement had produced a result that suggested that beauty quarks were decaying to muons less often than electrons, and what's more, the tension was just shy of the three-sigma threshold. When that first R_{K^*} result came out, Vava was immediately aware of the huge stakes. "It was clear to me back then that if the results were anything even close to correct, then with the full data set, we were sitting on something very explosive," he recalled. "We might be sitting on five-sigma new physics, so we better not muck it up!"

Vava and RX team's plan was to measure both R_K and R_{K^*} at the same time and in the process deal with any biasing effects that could affect both measurements. However, this choice introduced several complications. First of all, it led to tension with the RK team, who felt that the RX team was muscling in on their research area. It also raised the prospect that this second result might not agree with the first one. Now, while it's perfectly possible for a theoretical prediction and an experimental measurement to disagree, two experimental measurements of the same thing must agree with each other, within their uncertainties. There is only one real world after all, and because both teams were claiming to be measuring the real world, they had better get the same answer. In this case the two measurements weren't just measuring the same thing; they were doing it *with* *the same data.* So the results ought to land smack bang on top of each other.

As with previous measurements, the RX analysis was performed blind, the team allowed to look at the final results only after receiving a thumbs-up from their review committee. Like a NASA rocket launch, the unblinding process would unfold step by step, with checks at each stage that had to be passed before the team would be allowed to move on to the next. The last and most critical of these steps was to check if their measurement of R_K agreed with the value published a year earlier. This really was the moment of truth: If the two numbers lined up, then all would be clear to move to the final step, and perhaps a conclusive discovery of a new force of nature. But if not, it would send the teams into a world of pain as they attempted to unpick the source of the disagreement.

The result wasn't what anyone had hoped for. The check revealed that the new value of R_K differed from the already-published value to an uncomfortable degree—especially when considering that they ought to be more or less identical.

With that, the unblinding process immediately came to a grinding halt as the two rival teams dug into the data in search of the culprit.

The most benign explanation was that the disagreement was merely down to a bit of bad luck. While both groups had used the same data, they had cut it up in different ways, meaning that they didn't share all the same particle collisions. In other words, you wouldn't expect them to be in absolutely perfect agreement. The difference between the two measurements was rather large for such a random statistical wobble, but it wasn't beyond the realm of possibility. If that really was the explanation, then all would be clear to proceed with the final unblinding.

However, as the teams dug deeper, they began to find ominous signs that all was not well. The discrepancy between the two results seemed to come from the different ways the two studies had dealt with particles masquerading as electrons. Such fakers are generally referred to as "background," similar to the kind of interference you might hear on your car radio when a pirate station breaks through on the same frequency as the channel you're listening to. One particularly dangerous background for these measurements is when a beauty quark decays into a strange quark and two hadrons (a catchall term for particles made from quarks like protons and neutrons). Now, since electrons and hadrons interact with the detector rather differently, it should be relatively unlikely that you'll get them confused. However, the decays to hadrons are much more common than the decays to electrons, and since the detector isn't perfect, there will always be a certain fraction that slips through the net.

The way LHCb deals with this is by combining lots of information from its various subsystems to come up with an estimate of the likelihood that a given particle is an electron or a hadron, or indeed something else. We then impose some requirements on the data—for instance, "only accept particles that have a better than 90 percent chance of being electrons." If you set this threshold high enough, you can kill off the background from hadrons and ensure a pure sample of electron decays.

The RK and RX teams had used slightly different requirements to get rid of these hadron backgrounds, with the RX team imposing a stricter threshold for a particle to be counted as an electron. Now, in principle, this shouldn't have been a problem, because both teams had used simulated data to estimate how many hadron decays should sneak into their data sets and both had found it to be negligible. But the discrepancy now raised a frightening possibility: Had we all grossly underestimated how many hadrons were slipping through? Could our nets be much leakier than originally supposed? If so, then the apparent anomalies, the best hope for finally reaching beyond the standard model, could be nothing more than a mirage.

This awful possibility caused Mitesh "moments of absolute terror," as he put it, "proper bottom-falls-out-of-your-world terror. Oh my God, have we really done this wrong?"

With a mounting feeling of dread, I got into a frenetic back-and-forth with Paula from the RK team about how we might test for missed backgrounds. Realizing that a particular type of background would show up as a characteristic spike in a particular graph, I asked Paula if she had the graph to hand. She didn't, but within an hour she'd produced it from the data used in the 2021 analysis. When she sent it to me, my heart sank. There it was, an unmistakable spike just at the location you'd expect for missed backgrounds. Frustratingly, I had to log off to head to a doctor's appointment but told Paula I'd be back online that evening. She messaged back, "Might not be a good moment for the doctor to measure your heart rate."

Our fears abated somewhat over the next few weeks. First of all, Paula, Mitesh, and their team remeasured R_K, cutting out the bit of data with the background spike, and while the result did inch toward the standard model prediction, it moved by only a percent or 2, nowhere near enough for the background we'd found to be the sole cause of the anomaly. Meanwhile, the RX team had significantly tightened the net they used to filter out hadrons and found that the new value of R_K hardly budged at all. If missed hadrons had really been the cause of the anomaly, it should have shifted decisively toward the standard model.

Reassured that the backgrounds were under control after all, the RX team finally got permission to unblind their second result, the so-called R_{K^*} ratio. But rather than offer the discovery of a new force, the result threw us back into fear and confusion. The value of R_{K^*} had come out much higher than the result published in 2017 and, what's more, with both ratios in front of our eyes a clear pattern emerged: the values you got from the data recorded in 2011 and 2012 were consistently lower than those using the data taken from 2015 to 2018. Now, unless the laws of physics had miraculously changed in the intervening period, this pointed loud and clear to a big problem. With alarm bells ringing in all our heads, the RX team again imposed stricter requirements designed to reject hadrons and remeasured R_{K^*}. This time, the value jerked violently upward toward the standard model.

At long last it was clear: our measurements were contaminated.

The RX and RK teams worked furiously to understand how we had got the estimates of the amount of background so badly wrong. Meanwhile, rumors began to swirl outside LHCb that something was up with our anomalies. The collaboration was now faced with a dilemma: put out the result as fast as possible to let the world know there was a problem, but without a detailed understanding of what it was; or redo the analysis carefully and precisely and release the result only when we were sure it was solid, but with an inevitable delay.

The summer of 2022 leading into autumn was difficult for everyone involved as both teams worked themselves to the bone to solve the problem and prepare the results for publication. To avoid rumors filling an informational vacuum, LHCb agreed to attach a statement to our conference talks saying that we had gained a deeper understanding of the lepton universality measurements and that new results would arrive in the relatively near future. Vava, the head of the Paris team, described a particularly dark moment at the prestigious International Conference on High Energy Physics in July 2022, when the audience burst into laughter as the LHCb statement was read out.

"That was definitely a low point in my career," he told me, "because whatever else you do, you hope to not be a joke, and it was clear the laughter was genuine. We were saying in a very convoluted way, 'We fucked it up, lads.'"

For Vava, the lead-up to the moment when the result would be revealed to the wider world was especially tough. Alongside his work on RX, he was managing a major part of the upgrade work on the LHCb experiment. That July, a year later than expected, the third run of the Large Hadron Collider had begun, with Vava and his colleagues spending long hours in the control room getting the system into gear. Meanwhile, his father had been diagnosed with cancer and, although the prognosis had originally looked hopeful, at noon on October 27, just as they were about to test the system for the first time, his mother called to tell him his father had died suddenly. On the very same day that the project Vava had spent years working on produced its first signature of real particles, he had to leave the control room and drive a thousand kilometers to Vienna to be with his family. He recalled being at a gas station at 11:00 p.m. when an email came through with a graph showing that their system was working. "It was so surreal, the whole thing," he recalled. "To unroll all of that from the completion of the measurement is very, very difficult."

It was on December 20 that Renato Quagliani, the postdoctoral researcher who had done much of the work on the RX analysis, had the unenviable task of giving the particle physics world a very unwelcome Christmas present. Speaking in the main CERN auditorium, he gave a courageous account of the new analysis, including the huge amount of detailed work that had gone into understanding the backgrounds that had polluted earlier measurements.

But impressive as the huge body of work was, it couldn't cushion the punch to the gut that came when he showed the final results: With the backgrounds accounted for, both R_K and R_{K^*} agreed with the standard model prediction. The anomalies that had promised to transform physics had evaporated.

Where Are We Now?

"Aime la vérité, mais pardonne à l'erreur," wrote Voltaire. Love truth, but pardon error. The truth, at last, was out. But would the wider physics community pardon our error? As the dust was still settling on the bombshell dropped by the RX team, I caught up with some of the theorists who had devoted years of their recent careers to understanding the anomalies emerging from LHCb.

First up was Gino Isidori, the theorist from Zurich who had become one of the most enthusiastic advocates for the anomalies and their potential to lead to a more unified theory of physics. I had expected to meet a man dejected and perhaps angered by the recent turn of events, but to my surprise he remained defiantly optimistic.

"I admit we had a bit of an unpleasant Christmas," he said with a wry smile. "I don't know if LHCb planned it on purpose so we couldn't complain during conferences! But now I've passed through the more negative phase." The anomalies had inspired Isidori to work on models that unify quarks and leptons, explaining the origins of the different generations of matter and at the same time giving a more unified account of the fundamental forces. The loss of the lepton universality anomalies was clearly a big blow, but as far as Isidori was concerned, the model he's been working on is still very interesting. "There are a lot of colleagues who are super depressed, but I am not, because I am convinced that the model is good."

What's more, for Isidori, it was actually a different anomaly that first led him to explore his unification models. The anomalies in how beauty quarks turn into strange quarks might have faded away, but there are long-standing anomalies in a different set of processes where a beauty quark turns into a *charm* quark and either a tau or a muon (as a reminder, a tau is the third-generation version of the electron). Isidori describes these anomalies as "the mother of the anomalies," and for as long as they remain in play, they support his theory.

As for what he thought about how LHCb handled the whole debacle, he said, "I traveled to CERN to see the seminar before Christmas. I intervened saying that I was a bit disappointed with the way the experimentalists had accumulated positive evidence for a while and then there was this big drop. But still it was good that LHCb discovered the problem and not another experiment, so we cannot complain too much, it's part of life. It was honest that they didn't hide the problem and they went deeper trying to understand their own data."

Even if *all* the anomalies go away, Isidori still feels that they will have taught us something valuable. "We were lucky that these anomalies appeared because they led us to think about new ideas. Theories that unify quarks and leptons at low energy—nobody had ever thought about them before. Maybe we just have to work harder to find evidence."

Isidori is excited for new evidence that may be coming soon. In Japan, the Belle II experiment has been collecting data on a wide range of beauty quark decays. If they were to confirm the remaining anomalies at LHCb, that would totally change the game. That said, Isidori did admit that not all his colleagues feel the same way. "Some of them tell me I'm flogging a dead horse," he said. "And of course it's not the only thing I'm working on. But you have to stay optimistic."

Not everyone shares his rosy outlook. Ben Allanach is a theoretical physicist at Cambridge who has spent many years thinking about the LHCb anomalies. In contrast to Isidori's top-down unification project, he's been working from the ground up, proposing the addition of a single new force, called Z-prime, that pulls on electrons and muons with *different* strengths. However, these models were mostly motivated by the differences in beauty decays to strange quarks, and now that they're gone, his theory does look less viable.

"When I saw the results before Christmas, I thought, 'That's it, I'm downing tools. I'm gonna have to think next year about where my research is going to take me.' I was gutted because I was a believer. Now I'm not sure. Perhaps I got a bit too excited and carried away."

For Allanach the remaining anomalies are interesting, but now he thinks it's more like a fifty-fifty chance that they're bona fide signs of new particles. Along with the beauty to charm quark decays, there are still anomalies in the decays of beauty quarks to strange quarks and two muons. Almost every measurement of these processes shows they are happening less often than predicted by the standard model and that the angles that the particles emerge at are also different from what's expected.

"There's clearly something to be understood, but maybe it's just the standard model," Allanach continued. "I'm still working on it. It's stimulated some ideas that I still think are interesting, a paradigm that we've never explored before. I'm trying to look on the bright side."

Does he think the recent turn of events could spell trouble for particle physics in general?

"For the field as a whole it's a storm in a teacup. Look at the last hundred years. There have been plenty of moments like this. It may feel a bit cutting now, but it'll all come out in the wash."

I'd last spoken to Mitesh Patel at Imperial College London back when LHCb had been basking in the afterglow of a year of exciting results that seemed to all be pointing toward a major discovery. I had asked him how he'd react if things turned against the anomalies. "Well," he said, "even if it turns out we didn't discover new physics, at least we *might* have. It's better to have loved and lost." Speaking again in the aftermath of the body blow of the RX results, I wondered whether he remained philosophical about the whole thing.

"It hurt, there's no question," he said. "It's not nice to be exposed in that way on a big stage." However, it was clear that he wasn't prepared to wallow, nor to let his more junior colleagues slide into despair. "I've been trying to rally the troops, who've been more and more depressed by the experimental news," he said. "It's still the case that we're just one measurement away from turning this whole thing around. And while that's true there's still work for us to do."

I was surprised by his zeal. He spoke with a sense of mission that I'd found hard to conjure myself over the previous few months.

"What is our job as experimentalists? We've got this instrument, and we've got to use it. I want to break the standard model. That remains undiminished. Obviously, I wish we hadn't made a complete cock-up on the biggest stage possible. But I can't undo it, so what should I do from here?"

One new result from the CMS experiment, one of the two largest detectors at the Large Hadron Collider, generated a bit of hope in the summer of 2022. They had been searching for "leptoquarks," particles that can pull off the unique trick of decaying into a beauty quark and a tau at the same time. Leptoquarks are a key prediction of the unified models that Isidori and his colleagues have been working on, so seeing one created at the LHC would be a game-changing discovery. What CMS saw was the *hint* of such a particle. It was at about the 3.4-sigma level, again shy of a 5-sigma discovery, but nonetheless worth keeping an eye on.

As for what LHCb can do, Mitesh is now focused on new beauty quark decays that might give even greater sensitivity to the influence of new forces. These new decays are extremely hard to detect, but as Mitesh sees it, our job is to always be working at the limits of what's possible. "We have to be pushing the envelope, and if you're really genuinely pushing the envelope, you are taking enormous risks. Maybe we'll screw up, but if you want to do something genuinely new, that's the price you pay. It is the job to put your arse on the line.

"I would actually like to break the standard model," he reiterated, "and even if nobody else ever finds out, I still don't care. I want to see with my own eyes what nature's got to show us."

Vava, who led the RX team, shares Mitesh's sense of purpose: "It's like Pascal's wager; it's better to believe in new physics, search for it, and not find it than waste your career being pessimistic. So, let's do whatever needs to be done to give us the best possible chance of catching new physics if it's there. If you're going to be in this field at all, that has to be your approach."

Despite the setback of December 2022, there are still a plethora of puzzling results, for me and my fellow physicists, that could be pointing to the next revolution in our understanding of nature at the most fundamental level. Or they could all be red herrings. Either way, they will have left us with a legacy knowledge—from new theoretical paradigms to experimental techniques.

As Mitesh said, whenever you are working at the limits of knowledge, you run the risk of making mistakes, but errors can be the greatest training of all. When I was an undergraduate, one of my tutors, a white-bearded wire of a man named Bob Butcher, had a simple phrase stuck above his desk that has stayed with me: "I've learned so much from my mistakes, I think I'll make another."

Chapter 8

War in Heaven

A disagreement over the expansion of the universe is
dividing cosmology. Are we on the brink of a revolution
in our understanding of the universe?

On a bright Monday morning in July 2019, around thirty of the world's top cosmologists met in a small conference room at UC Santa Barbara to discuss an emerging crisis. Sitting in the front row was Adam Riess, a Nobel laureate who perhaps sensed that a second prize might be within his grasp. A professor at Johns Hopkins University and the Space Telescope Science Institute in Baltimore, he had already helped to rewrite the story of the cosmos once. Could lightning be about to strike twice?

In 2011, Riess had shared the Nobel Prize in Physics with Saul Perlmutter and Brian Schmidt for a discovery that had shocked the physics community—themselves included—and ushered in a new era in our understanding of the universe. For more than half a century, cosmologists had believed that after an initial rapid burst during the big bang, the expansion of space should have been slowing down as gravity fought to pull distant galaxies back toward one another. However, in 1998, their studies of distant stellar explosions, known as Type 1a supernovas, had revealed the complete opposite: the expansion of the universe wasn't slowing down; it was accelerating.

The implications were profound. It implied the existence of

a hitherto unknown form of energy permeating the cosmos whose repulsive influence overwhelms gravity at the grandest scales and causes the universe to inflate at an ever-increasing rate. They had uncovered the repulsive force known as dark energy, which to this day remains a mystery.

Following their momentous discovery, Riess put together a team of scientists with the aim of teasing out the nature of this mysterious substance. The project became known as SH0ES, which stands for "Supernovae, H_0, for the Equation of State of Dark Energy" and is pronounced like the things you wear on your feet. By tracking the expansion of the universe in exquisite detail, they hoped to reveal precisely how dark energy sculpts the cosmos, giving cosmologists clues to its true nature.

To achieve this, they had spent the preceding decade and a half making ever more precise measurements of the so-called Hubble constant, a single number of truly cosmic significance.

The Hubble constant tells us how fast space is expanding by relating how far away a galaxy is to how fast it's rushing away from us due to the expansion of space. More important, it also allows cosmologists to run the clock backward to the big bang and figure out the age of the universe.

Broadly speaking, there are two different approaches to measuring the Hubble constant. The first is to look up into the heavens at astronomical objects like stars and galaxies, then estimate both how far away they are and their speeds. The relationship between those two figures allows you to calculate the value of the Hubble constant; the more precise your estimates of distance and speed, the more precise your value. This method is often referred to as the direct approach, or sometimes the local approach. (Seeing as some of the galaxies used in this method can be more than ten billion light-years away, "local" might seem an odd term, but then again the observable universe is more than ninety billion light-years across, so I guess that counts as *relatively* local. Never trust a cosmologist if they suggest going for a short walk.)

The second method, sometimes called the indirect or early universe approach, does something rather more subtle. By studying the faded light of the big bang—known as the cosmic microwave background—cosmologists are able to deduce certain properties of the early universe, in particular the amounts of matter, radiation, dark matter, and dark energy it contained. Then, using the standard model of cosmology—the prevailing theory describing how the universe evolved, not to be confused with the standard model of particle physics that we've mostly discussed so far—they can take this information and *predict* how fast the universe ought to be expanding today.

The crucial point is that these two different methods—direct and indirect—should give us the same answer. The universe can't be expanding at two different rates at the same time.

The problem is, they don't. The SH0ES team's direct measurements of the Hubble constant were coming out significantly larger than cosmologists had predicted based on their understanding of the early universe and our current theory of cosmology. The direct and indirect measurements were in conflict. If the SH0ES team was right, then this growing rift—the Hubble tension, as it came to be known—threatened to rupture the standard model of cosmology and precipitate a crisis of a kind not seen for decades.

All this was the backdrop for the meeting convened that summer morning in Santa Barbara. Riess was well aware that many of the people in the room had serious doubts that the Hubble tension was anything more than the product of systematic biases in his team's measurements. Sitting just behind him in the second row was Wendy Freedman, one of the world's most respected astrophysicists, who had made her name with the first genuinely precise measurements of the Hubble constant during the 1990s. At the time, there had been a fierce disagreement between two rival teams, one claiming that the Hubble constant had a value of 50, the other 100 (we'll get to the units of these numbers and what they mean exactly in a little bit). Using the newly launched Hubble Space Telescope, Freedman and her team had decisively resolved the debate, getting a value of 72 with an uncertainty of about 8. In so doing, they ended decades of confusion and acrimony, fixing the age of the universe at around fourteen billion years.

Having devoted more than a decade of her life to measuring the Hubble constant, Freedman was well aware of the numerous pitfalls that awaited anyone attempting such a complex undertaking. Riess and his team were now gunning to shrink the uncertainty on the Hubble constant to a mere 1 percent, a level of precision unimaginable in the 1990s. Their hope was that this exquisite precision would finally allow them to break the standard cosmological model. But as would become clear over the following three days of presentations and discussions, Freedman was far from convinced that Riess and his team's measurements were solid enough to claim anywhere near the level of accuracy required.

Just a few weeks prior to the meeting in Santa Barbara, Riess and the SH0ES project had gotten an unexpected boost. Another team, an international collaboration calling itself H0LiCOW, had made their own direct measurement of the Hubble constant, only using a completely different type of astronomical object from that used by SH0ES. Strikingly, this independent measurement came out in near-perfect agreement with the SH0ES result.

Taking into account H0LiCOW's measurement, the disparity between the direct measurements and the early universe prediction had now grown to over five sigma—representing either an unambiguous discovery of new physics or a truly monumental cock up.

With the California sunshine streaming through the windows behind him, Riess took to the stage to kick off the proceedings, a spring in his step. He began by laying out the grand history of their shared field. "A century of cosmological research has led us to the standard model of cosmology," he proclaimed, before turning his focus to dark energy and dark matter.

Riess emphasized a glaring shortcoming of our best theory of the cosmos, one of which everyone in the room would have been painfully aware: we have no idea what most of the universe is made from. Between them, dark energy and dark matter make up 95 percent of the cosmos, and while astronomers have known about their existence for decades now, all attempts to discern their nature have come up empty-handed.

Given this massive gap in our knowledge, the current cosmological model contains a bunch of mostly blind assumptions about dark matter and dark energy. In particular, dark energy is assumed to be a form of energy that is inherent to empty space itself, with every cubic centimeter of the universe containing precisely the same amount, an amount that remains fixed throughout cosmic history. This notion of unchanging, uniform dark energy is known as the cosmological constant.

This is, however, just an assumption. It's more than possible that dark energy could be more exotic. Other possibilities include so-called quintessence or phantom energy, which, aside from sounding enticingly supernatural, are able to change over time or even vary from place to place.

As he opened his talk, Riess made the point that there was no reason to assume that dark energy was as simple as the current model states. Could the disagreement over the value of the Hubble constant from the early universe and from our more recent universe be telling us that dark energy has changed over cosmic history? It was a tantalizing possibility.

However, before cosmologists could begin to rewrite the standard model, Riess and his colleagues would have to convince themselves and everyone else that the Hubble tension was real and not merely a mistake in one of the measurements.

The SH0ES team's direct measurements of the expansion of space relied on stars known as Cepheid variables, whose brightness oscillates over time and which can be used to estimate distances to galaxies outside the Milky Way (much more on how this is done shortly). However, calibrating Cepheid stars is a tricky business. As relatively young stars, Cepheids tend to be found in the vast, star-forming dust lanes in central regions of spiral galaxies, and that dust has the troublesome effect of obscuring their light. A star that appears darker thanks to dust could be mistakenly assumed to be farther away than it really is. To be sure that the Hubble tension was real, other methods that didn't rely on Cepheids would be needed to cross-check the SH0ES result.

This was why the new measurement from H0LiCOW was particularly exciting. As Riess put it as he got into his stride, "I really want to draw attention to H0LiCOW because it has nothing to do with anything we've done. . . . This amazed me when I saw this last week."

The relationship between their findings was pretty remarkable. The SH0ES team's measurement of the Hubble constant had come out at 74.03 ± 1.42. That was already in conflict with the best prediction of the Hubble constant based on data from the early universe, which produced a figure of 67.4 ± 0.5. Meanwhile, H0LiCOW's direct method, which relied on observations of quasars—supermassive black holes that blast terrifically powerful beams of particles into space—resulted in a figure of 73.3 ± 1.8, in neat agreement with SH0ES.

After presenting his team's work alongside H0LiCOW's supporting evidence, Riess turned to another Nobel laureate sitting in the audience, the particle physicist David Gross.

"I know we've been calling this the Hubble *tension,* but we have a bona fide particle physicist in the room, David Gross, and they always tell us you got to get to five sigma, so are we allowed to call this a *problem* now?"

Gross paused for a moment to think. "In particle physics we wouldn't call this a tension or a problem, but a crisis."

Riess responded, "Okay, we're in crisis everybody!"

If the budding drama wasn't enough, the next day brought another unexpected turn of events. Just after lunch, Wendy Freedman got up to present a brand-new measurement of the Hubble constant of her own, a direct measurement, which she and her team had released that very morning.

"Our chairman said this would be a completely uncontroversial talk," she began with cheeky understatement. It proved to be anything but.

"What I'm going to describe this afternoon is a completely different, from the ground up, redetermination of the distance scale based on the tip of the red giant branch." In other words, Freedman and her team had used a different method, one that swapped out Cepheid variables for red giant stars. Having spent years working with Cepheids herself, she had serious doubts about their reliability, at least at the levels of precision

SH0ES was claiming. Red giant stars, on the other hand, could be found in the outer halos of galaxies, far from the obscuring influence of dust. She argued they offer a potentially cleaner route to measuring the Hubble constant.

Her team's measurement landed smack bang between SH0ES and the standard cosmological model prediction: 69.9 ± 1.9. What just a day earlier had appeared to be a clear-cut disagreement between the direct measurements and the early universe predictions had suddenly gotten a whole lot murkier.

During the Q&A that followed, Freedman and Riess engaged in a somewhat testy exchange as Riess questioned whether she had properly accounted for various effects. Freedman responded that he had misunderstood their methods. One member of the audience suggested that the two cosmological titans needed to "duke it out" to settle the matter. As she concluded her talk, Freedman magnanimously suggested she and Riess put their heads together outside the conference room.

The meeting had begun with what appeared to be a triumphant vindication of SH0ES and the Hubble tension. Now a more confusing picture was emerging. Clearly, both of them couldn't be right.

The next two days brought three *more* direct measurements of the Hubble constant, each based on different types of stellar objects. And in each case, they landed close to or even higher than the SH0ES result, bolstering Riess's case. The SH0ES team member Dan Scolnic pithily summarized the confusion in the room on Twitter: "This week is too much. Go home Hubble constant, you're drunk."

The conference ended with a poll on whether the situation was a curiosity, tension, problem, or crisis. While three-quarters of the room voted for either tension or problem, Riess alone put his hand up to vote for crisis. The delegates left on Wednesday afternoon with little clue where this strange saga would lead, but with plenty of work ahead of them.

So, how might this crisis be resolved? And what could it mean for our understanding of the universe? To answer these questions, we first have to delve into the precise nature of the Hubble tension, and in particular how the expansion of space is measured in practice. It is in these gory details that the devil may lie. Indeed, the origins of the Hubble tension go right back to the birth of modern cosmology at the start of the twentieth century. It was in those early years that astronomers first began to build the ladder that would allow them to reach out to the most distant parts of the heavens.

Climbing the Cosmic Distance Ladder

There's a famous scene in the Irish sitcom *Father Ted* where the eponymous protagonist tries in vain to explain the concept of perspective to his half-witted fellow priest, Dougal. As they sit together in a caravan on a rather dismal holiday, Ted looks at Dougal wearily. "Okay, one last time," he says, picking up two plastic model cows, "these are *small*," then gesturing to the window, "but the ones out there are *far away*." Dougal's brow furrows with effort, but he soon shakes his head and Ted throws the models to the table in frustration. Now, I'm not in any way suggesting that astronomers are as dim-witted as Dougal, but they suffer much the same problem when it comes to stars.

Measuring the distance to a star or galaxy is *hard*. I'm sure at some point you'll have been out at night and noticed a bright light in the sky and been briefly confused as to whether it's a star or planet, or maybe just an approaching plane. Usually, you can tell it's a plane from whether it appears to be moving—obvious for the simple fact that a plane is far, far closer to us than even the closest astronomical body. But when it comes to estimating the distances to stars or galaxies, the task is much more difficult.

The question of how far away stuff is in space is central to measuring the Hubble constant, a figure that, to remind readers, describes the relationship between how fast a galaxy is rushing away from us due to the expansion of the universe and how far away it is. Visualizing the expansion of space in three dimensions is famously mind-bending, but the best analogy I've come across is to drop down to just two dimensions. Imagine the universe as the surface of a balloon, with little galaxies drawn on in marker pen. As you blow air into the balloon, each galactic doodle gets farther away from every other, and the speed they move apart gets bigger the farther they are apart. This relationship between speed and distance is captured by an equation known as Hubble's law,* which tells us that there's a linear relationship between speed and distance; in other words, a galaxy twice as far away recedes twice as fast.

Now, someone once told Stephen Hawking that every equation added to a popular science book halves its sales. That may be, but I've rather made a rod for my back by writing a *whole chapter* about a parameter in an equation, so I kinda need to show you Hubble's law now. (If the sight of mathematics gives you horrible high school flashbacks, consider this a trigger warning.)

---

* Sometimes also referred to as the Hubble-Lemaître law to acknowledge the important theoretical work of the Belgian astronomer Georges Lemaître on the expansion of the universe.

$$v = H_0 D$$

This is just about as simple an equation as you could hope to encounter, having only three parameters: v represents the speed a galaxy appears to be retreating due to cosmic expansion, D is its distance, and H_0 is the all-important Hubble constant. Since the Hubble constant allows astronomers to calculate the speed of a galaxy based on its distance, correspondingly, if you want to measure the Hubble constant, you need to measure both the speeds *and* the distances of a whole bunch of galaxies—the more the better. Plonk them on a graph and determine the gradient. The steeper the gradient, the bigger the Hubble constant and thus the faster the universe is expanding.

Fortunately, getting at a galaxy's speed is relatively straightforward thanks to quantum mechanics' greatest gift to astronomy, the fact that every element in the periodic table emits and absorbs characteristic wavelengths of light. These wavelengths correspond to photons being absorbed and emitted as electrons jump between different energy levels in atoms, and assuming that the laws of physics are the same throughout the universe (which they appear to be), these wavelengths are the same for an atom on Earth as they are for an atom in the atmosphere of a star on the other side of the universe. That means, if you break starlight into a rainbow spectrum, you'll see that it's scattered with dark and bright lines—a sort of stellar barcode. That barcode can then be compared with the known spectra of atoms measured in earthbound laboratories.

Armed with a galaxy's spectrum, astronomers can get at its speed. As photons travel toward us from a distant galaxy, they get stretched by the expansion of space, shifting the characteristic wavelengths emitted by atoms toward longer wavelengths. The longest wavelength of visible light is red, so light waves that get stretched due to the expansion of the universe are said to have been redshifted. Conversely, light that's been compressed to shorter wavelengths is said to have been blueshifted. By comparing the spectrum of redshifted light from a distant galaxy with the characteristic wavelengths measured in a lab here on Earth, we can calculate how fast that galaxy appears to be retreating from us.*

That's measuring speed. Measuring distance, however, is a whole lot trickier. A little like Dougal's cows, there is no immediately apparent way of knowing whether an astronomical body like a star is bright but far away or dim but close. Indeed, the question of how far away stuff is has exercised astronomers since time immemorial. Ancient thinkers like Aristarchus, Archimedes, and Ptolemy all had a crack at estimating distances to astronomical bodies such as the Sun and Moon. While their calculations were based on sound geometry, they mostly got answers that were way off thanks to the near impossibility of making precise measurements before the invention of the telescope. Many of these early attempts were based on parallax, which may sound fancy and mathematical but is actually the phenomenon most of us use every day to judge distances. Par-

---

* There are a couple of subtleties here. First comes from the fact that a galaxy can be moving around relative to us, in addition to its motion due to the expansion of space. Imagine a person walking away from you down an escalator, where the escalator represents the expansion of space. This can add further redshift or blueshift, depending on whether the star is moving away from or toward us when you subtract the effect of cosmic expansion. Another complication comes from the fact that as a photon climbs out of a gravitational field, say the gravitational field of a galaxy, it also gets redshifted. So astronomers need to take both effects into account when calculating the speed of expansion. It is, suffice it to say, no straightforward task.

allax is the simple geometrical fact that objects at different distances appear to shift their positions when we change where we look at them from. So, for example, if you hold out your arm and extend your thumb and line it up with say a door frame or a window, and then look at it through just your right and then just your left eye, you'll see that your thumb shifts to the left and then to the right. Our brains use the fact that each of our eyes sees a slightly different image to judge depth and estimate how far away stuff is. Likewise, pigeons, which have eyes in the sides of their heads with two fields of view that don't overlap, bob their heads up and down to achieve the same effect.

Mathematically, if you know the distance between the two different viewpoints (in the example I just gave, that's the distance between your eyes) and the angle that your thumb appears to shift by when you switch eyes, then you can calculate how far away your thumb is using just high school trigonometry. This makes parallax the gold standard for measuring distances to objects you can't actually reach with a measuring tape. However, the problem with parallax is that the farther away an object is, the less it shifts. If you try the opening and closing one eye trick, but this time swapping your thumb for a distant tree or building and comparing it with the far horizon, you'll be hard pressed to notice a shift at all. Stars are so stupendously far away that you have absolutely no chance of seeing parallax with your naked eyes. However, if you increase the distance between your two viewpoints and use a very accurate telescope, then in principle you can make the shift large enough to notice. Handily our solar system provides us with a useful baseline: Earth's orbit around the Sun.

The diameter of Earth's orbit is approximately 300 million kilometers (roughly four trillion times the distance between your eyes). So if you have a telescope that can image the night sky with superb precision, then photographs taken six months apart will reveal tiny shifts in the positions of nearby stars relative to those farther away. Measuring so-called stellar parallax only became possible in the nineteenth century thanks to improvements in telescope technology, allowing the Scottish astronomer Thomas Henderson to estimate the distance to Alpha Centauri as 3.25 light-years, using observations he made in 1832 and 1833.[*]

However, parallax has its limitations. Even using the most advanced space-based telescopes available today, parallax can only effectively estimate distances to stars that are within a few tens of thousands of light-years of Earth. The Milky Way, to put that in context, is roughly 100,000 light-years across. This means that parallax is unable to reach most stars within our own galaxy, let alone other galaxies. However, there are ways to extend distance measurements to more remote parts of the cosmos.

We owe our ability to measure distances to objects outside our own galaxy to the American astronomer Henrietta Swan Leavitt. Leavitt began her career at the end of the nineteenth century, a time when few positions were available to women in astronomy. It was while she was working at the Harvard Observatory cataloging Cepheid variable stars that she made a discovery, one that allowed astronomers to build a measuring tape that could stretch far beyond the confines of our galaxy.

Cepheids are young, giant stars with pulsating atmospheres that grow and shrink periodically over time, creating a regular heartbeat-like oscillation in how bright they appear. Leavitt's discovery was that there was a clear relationship between how

---

[*] Henderson's measurement is about 25 percent smaller than the modern figure, but it's an impressive estimate nonetheless.

fast a Cepheid pulses and its average brightness; the brighter it appears, the slower it pulses. She realized that this meant Cepheids might serve as "standard candles" (a term she coined), allowing astronomers to determine distances wherever Cepheids were visible.

Here's how it works. As discussed, the problem with determining distances to stars is that not all stars are the same: some are larger and more luminous, others smaller and dimmer. How bright a star appears from Earth depends on how much light it pumps out and how far away it is, with more distant stars appearing dimmer. To be precise, how bright a star appears falls off with its distance squared, meaning that a star twice as far away appears four times dimmer.

A "standard candle," on the other hand, is a type of star where you *know* the star's total light output. Since its brightness is known, you can work out how far away it is from us by how bright it *appears* on Earth.

Leavitt's discovery that the speed a Cepheid pulsed could be used to figure out how much light it gives off was exciting because measuring how fast a Cepheid pulses is relatively easy; you just need to watch how long it takes to brighten and dim, which can be anything from a day to a few months. From its pulse period, you can deduce its light output and thus how far away it is.

However, for this approach to work, you first have to determine the exact relationship between light output and pulse period for a whole bunch of Cepheids. This is where parallax comes to the rescue. There are, luckily for us, several Cepheid stars in our own galaxy that are close enough that we can make direct parallax measurements of them. And once you know how far away a star is based on parallax, you also know how much light it's giving off based on how bright it appears on

Earth. By measuring the distances to enough Cepheids this way, you can then accurately calibrate the relationship between pulse period and light output and, hey, presto, you have a cosmic measuring tape.

Leavitt's law, as it has become known, revolutionized our understanding of the cosmos. At the start of the twentieth century, astronomers believed that the Milky Way was the entire universe: a giant island of stars alone in the darkness. However, a debate had raged over whether "spiral nebulae"—faint whirlpool-like smudges in the night sky—were clouds of dust and gas within the Milky Way, or perhaps galaxies in their own right. In 1923, the American astronomer Edwin Hubble discovered a Cepheid in the largest of these nebulae, Andromeda, and using Leavitt's law he estimated that it was around a million light-years away, a distance that was a thousand times larger than the established size of the Milky Way at the time. Suddenly the known universe had ballooned, expanding from a single island universe to one populated with untold distant galaxies.

Leavitt's work would soon have even more profound effects. In 1929, Hubble used Leavitt's law to estimate distances to twenty-four nearby galaxies containing Cepheid stars. When he compared these distances with measurements of how fast these galaxies were moving made by his fellow American astronomer Vesto Slipher, he noticed a striking relationship: the farther away a galaxy was, the faster it appeared to be retreating. This led to the extraordinary realization that space itself was expanding—the aforementioned Hubble's law—and ultimately to the big bang theory. Leavitt's pulsing stars had not only extended the boundaries of the cosmos; they had revealed that it had a beginning.

However, it would soon become clear that Hubble's own

measurements of distances were on shaky ground. In 1931, Hubble and his colleague Milton L. Humason published the first determination of the Hubble constant, arriving at a value of 558 kilometers per second per megaparsec (a megaparsec is around 3.26 million light-years). What this value means is that a galaxy 1 megaparsec from Earth should be retreating from us at 558 kilometers per second. It didn't take long for cosmologists to realize that Hubble and Humason's result posed a serious problem: the value of the Hubble constant they'd arrived at was so large that it implied the universe was a mere two billion years old. Geologists had discovered rocks that were significantly older than that. Clearly the Earth couldn't be older than the universe itself. For opponents of the big bang theory, this became a key piece of evidence against the idea that the universe had a beginning.

However, in 1952, the German astronomer Walter Baade realized that Hubble and Humason's calculation contained a serious flaw. Unbeknownst to them at the time, the Cepheids they had used in their calculation actually came from two distinct populations of stars: one group was young and rich in heavy elements (referred to as metal rich), while the other was older and relatively metal poor. Crucially, the relationship between the pulse period and light output for these two distinct groups was different. When Baade accounted for this, the size of the Hubble constant crashed to just half its previous value, which had the rather dramatic knock-on effect of doubling the size and age of the universe.

In the subsequent decades, new measurements sent the Hubble constant tumbling further, as astronomers discovered various troublesome effects, including that many of the brightest Cepheids that Hubble had observed were in fact groups of overlapping stars, or even clouds of gas lit by starlight.

By the start of the 1990s, the situation had reached something of an impasse, with two rival factions arguing for values of around 50 kilometers per second per megaparsec and 100, respectively. It was becoming increasingly clear that ground-based telescopes simply couldn't see far enough into space to settle the issue. To spy Cepheids at even greater distances would require a telescope that could peer deep into space from above the hazy veil of Earth's atmosphere. Enter the Hubble Space Telescope.

The Hubble Space Telescope is most famous for its stunning images of distant galaxies and towering pillars of gas and dust. However, it has done far more for astronomy than produce pretty pictures for Neil deGrasse Tyson to stand in front of while looking dreamy-eyed. When Hubble was first launched in 1990, one of its key aims was to measure the Hubble constant with a precision of 10 percent and to finally settle the debate over this unruly cosmic parameter.

Led by Wendy Freedman, the Hubble Space Telescope Key Project, as it became known, first aimed to extend the range of Cepheid observations, discover hundreds of new pulsating stars, and measure distances to a larger number of galaxies. In 1994, she and her team announced their first result. They had arrived at a new measurement of the Hubble constant: 80, roughly halfway between the two contested values.

It was a promising start, but Freedman's team hadn't yet reached their target precision of 10 percent. To get there, Cepheids alone wouldn't be enough. Astronomers would need to reach out to what is called the Hubble flow, the region where the expansion of space becomes so rapid that it completely dominates the motion of galaxies. This meant reaching incredibly remote galaxies, hundreds of millions of light-years away. Beyond distances of about 150 million light-years, however,

Cepheids are too faint to observe reliably, even with the Hubble Space Telescope. At such immense distances, only extremely bright objects can be used as standard candles. Fortunately, nature has obligingly provided us with one: cataclysmic stellar explosions known as Type Ia supernovas.

Type Ia supernovas occur when a white dwarf—the glowing husk of a dead star about the size of Earth—sucks material off a companion star. Eventually, the crushing weight of the pilfered star stuff accumulating on the white dwarf's surface causes its temperature to rise so high that nuclear fusion restarts in its core, causing the white dwarf to detonate violently, utterly obliterating the star and releasing an awesome blast of light and matter. These explosions briefly shine brighter than a billion suns and can be seen at vast distances. Crucially, astronomers believe that Type Ia supernovas all pump roughly the same amount of light into space, making them another standard candle (or perhaps more accurately a standard firework). If you can figure out how luminous Type Ia supernovas are and spy them in a number of distant galaxies, you can work out how far away those galaxies are from how bright the supernovas appear.

Cepheids are found in almost all galaxies. Type Ia supernovas, on the other hand, are exceptionally rare, occurring only about once every five hundred years in the Milky Way. That means that if you wanted to calibrate the light output of Type Ia supernovas using parallax, you'd be waiting around for millennia. If you have any friends in academia, you'll know that research contracts tend to be rather shorter than that. So instead, Wendy Freedman's Key Project team searched for Type Ia supernovas in other galaxies, beyond the Milky Way but still close enough to measure their distances using Cepheids. These distances then allow you to measure the true light output of a supernova. The more such galaxies you find, the more accurately you can calibrate them.

Once you've calibrated your supernova, you can then search for them in much more distant galaxies out in the Hubble flow, and from their apparent brightness you can determine their distances. Finally, armed with their distances and their redshifts, you can determine the Hubble constant.

This ingenious if roundabout method of tying different measuring techniques together is known as the cosmic distance ladder. Its most precise form has three rungs, the first based on parallax, then Cepheids, and finally Type 1a supernovas. To summarize it again in brief:

The Cosmic Distance Ladder

Rung 1: Determine distances to Cepheid variable stars in the Milky Way using parallax and hence calibrate how their pulse period relates to their light output.

Rung 2: Search for nearby galaxies that contain both Cepheids and Type 1a supernovas and use the relationship from rung 1 to measure their distances, and hence the light output of the Type 1a supernovas.

Rung 3: Search for distant galaxies in the Hubble flow with Type 1a supernovas. Use their measured brightness and their light output estimated in rung 2 to figure out how far away they are.

I am left somewhat in awe by the ingenuity involved in assembling the cosmic distance ladder. It's an incredible achievement, the product of two centuries of work by hundreds of astronomers that has utterly revolutionized how we think about the universe. In 2001, the Key Project team led by Wendy Freedman published their final tour de force measurement of the Hubble constant. Along the way, they had discovered hundreds of new Cepheids, measured distances to dozens of new galaxies, and combined multiple variants of the distance ladder to reach their targeted 10 percent precision.

The final result came out at 72 kilometers per second per megaparsec, with an error of just 8. Finally, after more than seventy years, the controversy over the Hubble constant had been settled.

Freedman later told me that she didn't think she'd ever work on the Hubble constant again. "I mean, why would I?" she said. The debate over its value had been settled once and for all. It was neither 100, nor 50, but 72. The universe's age (13.8 billion years) and size (90 billion light-years across) had been tied down, its rate of expansion determined. Job done.

Or so it seemed. Just as Freedman and her colleagues were moving on to new projects, two new teams were gearing up to enter the Hubble-constant industry, ready to blow the whole thing wide open once again.

The Once and Future Laureate

Adam Riess is a difficult man to pin down. As one of the world's leading astrophysicists, not to mention a Nobel Prize winner, he's in high demand. As I pulled up in the parking lot of the Space Telescope Science Institute in Baltimore, where Riess is based, on a sunny July morning, I had a lingering worry that after crossing the Atlantic, I might find he'd skipped town to attend a last-minute meeting. Reassuringly as I walked toward his office building, I noticed a car parked in the space closest to the entrance marked "Reserved for Nobel Laureate." Clearly the prize comes with ancillary perks.

As we saw earlier, Riess shared the 2011 Nobel Prize in Physics with Brian Schmidt and Saul Perlmutter for a discovery of the accelerating expansion of the universe. Their discovery had far-reaching implications, rewriting the universe's history and transforming our predictions of its ultimate fate. What's more, it led to the establishment of today's standard cosmological model, ending a long period of confusion about the makeup and evolution of the universe. It is thanks to Riess, Schmidt, Perlmutter, and their respective teams that we know of the existence of dark energy, the mysterious substance that drives the accelerating expansion of space and makes up around 68 percent of the total energy content of the cosmos.

Having helped to establish the standard cosmological model in the late 1990s, Riess, through his work with the SH0ES team, is now presenting it with the greatest challenge it has ever faced. Three years after the landmark meeting in Santa Barbara that heralded the start of the Hubble crisis, I had come to Baltimore to better understand how things had evolved since then, and to get Riess's unvarnished view of the current state of play.

After failing to find any sign of an official entrance, I let myself into the Space Telescope Science Institute via an unlocked side door, feeling more than a little worried that I was trespassing. The building appeared to be mostly deserted, though, being a Monday morning at the height of summer. I eventually found Riess's office on a dimly lit corridor on the ground floor. Much to my relief, he was waiting for me at his desk, an imposing wall of books and papers behind him. On a shelf near the door, I noticed a beach ball covered with an image of the cosmic microwave background, and alongside it his gold Nobel medal, glinting in the soft light.

To begin, Riess shared the origin story of his interest in the Hubble constant, which began in the aftermath of his Nobel-winning discovery. "My previous background is obviously on the accelerating universe and dark energy, but as a related side-line I was working on Hubble constant measurements. I was convinced that there was room for improvement."

As we've seen, direct measurements of the Hubble constant are made using the cosmic distance ladder. But the metaphor of a single ladder with multiple rungs is a little bit misleading. In fact, the cosmic distance ladder is more like three different ladders strapped together, hopefully rather securely, but with the potential for some wobbles at the joins. It was a desire to firm up this potentially precarious construction that led Riess and his colleagues to the Hubble tension in the first place. As he explained, "I felt like the field was ripe for improvement because the previous generation of Hubble Space Telescope instrumentation was not as good as the newer generations that were installed in 2002."

Unique among space telescopes, the Hubble, launched in 1990, was designed to be serviced by astronauts, who used NASA space shuttles to visit the telescope in low-Earth orbit every few years. This ended up saving the entire mission from disaster when astronomers discovered to their horror that the telescope's main mirror had been ground to slightly the wrong shape, resulting in blurry images that made the planned cosmology program more or less impossible. As the Hubble Space Telescope, and by extension NASA, became a subject of public ridicule and political ire, NASA engineers worked feverishly to come up with a way to fix the mirror. Salvation came during the first service mission in 1993, when astronauts fitted what effectively amounted to spectacles, correcting the telescope's fuzzy vision. Subsequent missions in 1997, 1999, and 2002 replaced or upgraded many of Hubble's instruments, all of which meant that by the early years of the twenty-first century the Hubble was far more powerful than it had been when Wendy Freedman and the Key Project team made their measurement, the one that had resulted in a Hubble constant of 72.

These upgrades presented Riess and his colleagues with the tantalizing prospect of finding even more distant Cepheid stars, allowing them to measure distances to a larger number of Type Ia supernovas, and hopefully pushing the uncertainty on the Hubble constant down yet further.

However, what really set a fire under the project was the prospect of a European Space Agency mission that promised to attack the Hubble constant from a completely different direction. The Planck spacecraft was designed to map the faded light of the big bang, the so-called cosmic microwave background, with unheralded precision, giving cosmologists a wealth of new information about the early universe. One output of Planck would be a new indirect measurement of the Hubble constant.

Here, Riess was at pains to make clear that Planck doesn't really measure the Hubble constant per se. Rather it provides input to a prediction. "The Hubble constant, of course, is a measure of *today's* expansion rate. So, you know, by definition you can't measure today's expansion rate shortly after the big bang; you can only predict it. Okay? So, they were going to make a very precise *prediction*, and it would be an important test of the cosmological model."

What was particularly exciting for Riess and his colleagues was that Planck's prediction of the Hubble constant would make it possible to check whether dark energy really was constant throughout space and time (consistent with the notion of a cosmological constant) or if it was something more exotic. Regardless, it was clear that Planck's prediction would end up being far more precise than the 10 percent uncertainty achieved by the Hubble Key Project. To make a serious test of the cosmological model, astronomers working on local measurements of stars and galaxies were going to have to up their game.

So, in 2005, Riess set up SH0ES, whose goal was to use the souped-up Hubble Space Telescope to push down the uncertainty on the Hubble constant, compare their more precise value to Planck's prediction, and tease out some new information about the nature of dark energy. To get there, the team would have to work hard to firm up some potentially wobbly rungs on the distance ladder.

First, the upgraded Hubble allowed the SH0ES team to deal with another potential spoiler: dust. The bane of many an astronomer, galaxies are threaded by huge dust lanes that obscure light from their stars, particularly near the galactic core, where most Cepheids are found. A new infrared telescope fitted to Hubble allowed it to peer through these dust clouds and get better measurements of Cepheid brightness. Second, and most important, the team used Hubble's improved resolution to discover new Cepheids in galaxies that had also experienced Type 1a supernovas. Previous measurements had been severely limited by being able to study only a small handful of such galaxies, but the SH0ES team's eventual aim was to boost the number to around forty. Such a large sample would remove the biggest roadblock to a more precise measurement and perhaps one day push the uncertainty on the Hubble constant down to a mere 1 percent.

"That was the road we set out on," Riess told me. "In 2009 we get to 5 percent uncertainty; 2011 we're at 4 percent uncertainty." At that point they had measured a value of 73.8 ± 2.4, which was consistent with the best cosmic microwave background prediction available at the time. So, as Riess put it, "it was fine, no big deal." Then, in 2013, ESA's Planck spacecraft released its first indirect measurement (or "prediction," as Riess would put it), based on its survey of the cosmic microwave background. Their value came out far lower than SH0ES's, at 67.3 ± 1.2.

At a stroke, the game had changed. Planck and SH0ES were in 2.5-sigma tension. This meant that while there was a chance that the results were consistent with each other, it was small: less than 1 percent. It was still a long way from 5-sigma, but it definitely looked as if *something* was up.

Suddenly Riess and the SH0ES team found themselves under the critical glare of the astronomical community. "Measuring the Hubble constant has always been really difficult. 'Could your measurement be wrong?'" he recalled hearing from colleagues. "But these weren't the days of photographic plates held up to the light trying to figure out how bright things were. We were using extremely modern methods and had really factored in systematic uncertainties and had redesigned the distance ladder to defeat them."

What gave Riess even more confidence in their result was that just a year earlier Wendy Freedman and her team had recalibrated their last measurement from the Hubble Space Telescope Key Project and got an answer that was consistent with SH0ES's, but not with Planck's. "So, in some sense we had already been cross-checked," he told me.

Nonetheless, it was immediately clear that a lot more work was going to be needed to convince the doubters. "I'd been through the accelerating universe and dark energy, so I know that if you get the answer everyone expects, you can have a five-page paper, but if you get the answer people *don't expect,* get ready for the scrutiny!

"So we did a big deep dive, and in 2016 we were like, we really have to figure out if we can do better." As before, the SH0ES team continued to expand the sample of galaxies with both Cepheids and supernovas, further securing the critical link between the second and the third steps of the ladder. However, this time they also targeted the join between the first and second rungs by using the Hubble to perform parallax measurements to nearby Cepheids in the Milky Way, something that had never been done before.[*]

"That 2016 paper was a big watershed moment for us," Riess said. This time, he and the team performed multiple different variants of the analysis, switching out different rungs for one another to see if any particular step could be the cause of the tension. But none of their different measurements got them anywhere close to the Planck value. The divergence between the two was becoming harder and harder to reconcile.

Their final result once again came out high, at 73.24 ± 1.74. In the meantime, Planck had released a more precise result of

---

[*] Time on the Hubble Space Telescope is in high demand, and since there are plenty of other telescopes powerful enough to study Milky Way Cepheids, Hubble had previously only been used to study Cepheids in galaxies too far away to be reached by other telescopes. Meanwhile, less powerful ground-based telescopes were used for Milky Way parallax measurements. However, different telescopes have different sensitivities to light (a bit like two computer screens with their brightness set to different levels), which could introduce a bias when comparing Cepheids in the Milky Way with those in distant galaxies. Using Hubble to do parallax measurements *and* observe Cepheids in other galaxies completely eliminated this problem.

their own, and it was even lower than before: 66.93 ± 0.62. The tension had now grown even stronger, reaching 3.5 sigma.

The next big step forward came in March 2019, when Riess and the SH0ES team used the Hubble Space Telescope to measure the brightness of Cepheids in the Large Magellanic Cloud, a dwarf galaxy in orbit around the Milky Way that can be seen on particularly dark nights as a hazy smudge in the Southern Hemisphere. Combined with another team's measurement of the distance to the dwarf galaxy, led by Grzegorz Pietrzynski in Warsaw, this resulted in a new value of 74.03 ± 1.42, increasing the tension even further to 4.4 sigma. They were on the brink of a discovery.

Then came the pivotal meeting at Santa Barbara, where H0LiCOW presented their own measurement of the Hubble constant using a completely independent technique. It was this that pushed the combined tension to over the symbolic five-sigma threshold for the first time. The anomaly had now gone way beyond the point where it could be simply explained away as a statistical fluke.

But it was also in Santa Barbara, as we saw, that Wendy Freedman dramatically burst back onto the Hubble scene with her own measurement, using red giant stars, a value that came out awkwardly *in between* the SH0ES and the Planck.

After Riess and Freedman's exchange, both teams got to work on honing their techniques and responding to criticism leveled at them during the conference. Freedman and her team struck first, offering a recalibration of their measurement based on red giant stars. It came out in perfect agreement with the result she'd presented in California. In further bad news for Hubble tension fans, in July 2020 a different team announced that they had reexamined the H0LiCOW analysis and found that they had made a flawed assumption. Once they made the necessary correction, their work sent H0LiCOW's measurement tumbling toward the Planck value, blowing up its uncertainty in the process.

Not to be outdone, Riess and SH0ES responded with two more papers of their own. The first landed in December 2020. It now included state-of-the-art parallax measurements to Cepheids in the Milky Way, performed by the European Space Agency's Gaia spacecraft, producing a result only a shade lower than their 2019 result and with a smaller error. Then, in December 2021, SH0ES put out their ultimate kitchen-sink measurement, a seventy-page monster paper that combined three different ways of calibrating Cepheid stars, forty-two examples of galaxies hosting both Type 1a supernovas and Cepheids, and seventy different variants of the analysis method. After a gargantuan effort, they had pushed the uncertainty on the Hubble constant down to a mere 1 (again, this is in kilometers per second per megaparsec). Most important, their measurement *alone* was now in five-sigma tension with the early universe measurement by Planck.

Riess seemed convinced that he had answered his critics as comprehensively as possible. More than that, the fact that every single measurement of the Hubble constant came out higher than Planck's tells him that there must be something going on.

"This now gets to a philosophical point," he said. "If this was just a measurement I'd be like, 'Oh my gosh, we've beaten this dead horse to a tenderized level!' If you look at the totality of all the local measurements, they're all high; the lowest is about 70. Let's say one hypothesis is that we're not doing great experiments. Why aren't we scattering on both sides? Why aren't some coming out below Planck?"

He continued, getting increasingly animated, "The bottom

line is not only are we at five sigma, but this is a *mature* measurement; it's been going on for at least ten years. When there are silly mistakes, they usually get cleared up in a year or two. I've never seen something that lingers longer. And other people have reproduced the measurements at every step."

A response from the wider community—one that clearly frustrates Riess—is that since their measurements disagree with the accepted standard cosmological model, they must simply be wrong.

"I've seen many people in the field being affected by their beliefs of what things *ought* to be. People say 'extraordinary claims require extraordinary evidence.' Is it really extraordinary to say that the standard cosmological model is wrong? We don't know what dark matter and dark energy are, we don't really know how many neutrinos there are, we don't have quantum gravity. If you look at it that way, then it's not an extraordinary claim at all."

Riess is now convinced that the SH0ES team's measurements are solid. As far as he's concerned, no one has been able to find a missed bias that could get anywhere close to bridging the chasm between their measurements and the prediction based on the standard cosmological model. What's more, the vigor with which he spoke gave me the feeling that he senses another major breakthrough is within his grasp—perhaps even that second Nobel. Toward the end of our conversation, he took me back to his student days for a salutary tale of the dangers of prejudging the outcome of your experiments.

"We used to have this great quantum physics professor. We'd come in the morning and tell him about some new experiment, and he'd say, 'It's wrong.' And we'd go, 'Oh, do you know about it?' and he'd say no. And we'd say, 'How do you know it's wrong then?' and he goes, 'Because they're all wrong.' Okay, sure, 99 percent of the time it turns out to be wrong. He was just making the smart bet. And that's fine if you're talking to your colleagues at a cocktail party and want to take a guess. But if you want to actually do the measurements, then that becomes a very dangerous attitude because you have no chance of measuring anything if you decide a priori that everything is wrong. It doesn't mean you don't have to be careful; you need to be careful. It doesn't mean you don't have to be critical; you definitely need to be critical. But you have to check your prior assumptions because if you have an expectation for what your experiment should find, you'll keep futzing with it until you get the result you expect, and that is really dangerous."

Before I left, I asked Riess how he thought the current situation compared with the late 1990s, when he and his colleagues were getting the first evidence that the universe's expansion was accelerating.

"It's a great question! In some ways they are totally opposite; back then our data was much worse, we had spent much less time at it, and we were like, 'We see this,' and after a short period of wandering in the wilderness, people were pretty quickly like, 'Sure, it's the cosmological constant, why not?' Then the cosmic microwave background people were like, 'We see it too,' and that was done.

"Here the data is much better than it was, we've been at this much longer, this one is easier to swallow. But the difference is we don't have a name for it. Back then we had 'dark energy,' even if we didn't have any real idea of what that was. Whereas now we haven't zeroed in on the most likely suspect. It's not my mindset, but I know people who say, 'I won't accept any result that doesn't have a theoretical explanation,' which is a funny way to look at things, I think.

"I would love to have someone give me a story about how we can solve this, but no one has so far."

The Master Chess Player

"I've been at this for a while," Wendy Freedman told me over a video call from her office, a picturesque view of the University of Chicago's campus over her shoulder. She started her career in the early 1980s, back when astronomers were still using photographic plates. Based on technology developed in the nineteenth century, plates were fiendishly difficult to work with: they didn't respond to light in a uniform way and were mostly sensitive to the blue end of the spectrum, which meant dealing with dust at the red end was nearly impossible. Dust, she reminded me, scatters the light from Cepheids, making them appear dimmer and farther away than they actually are. And if you're trying to measure the Hubble constant, that's a big problem.

Those were the days when rival teams claimed values that disagreed by a factor of two: one getting 50, the other 100. It was only through the launch of the Hubble Space Telescope that Freedman and the Key Project team had the opportunity to settle the controversy. That seminal work made her career and established her as one of the world's most respected astrophysicists. It also gave her a deep understanding of Cepheid stars and the challenges of constructing a cosmic distance ladder.

It's this deep knowledge that underlies her skepticism about today's Hubble tension, where cosmologists are now arguing over effects down at the level of a few percent.

She recalled an adage from Allan Sandage, a former colleague at the Carnegie Observatories in Pasadena, California, and scientific heir of Edwin Hubble—who had spent most of his career working on the expansion of the universe. He once told her, "Measuring the Hubble constant is like being a master chess player, and only a master chess player can play the game, because only they know how all the moves on the board fit together."

If by implication Sandage was claiming to be one such master player, then Freedman undoubtedly is too. Sandage had long argued for a low value of the Hubble constant, leading to that value of around 50. Freedman ultimately proved him wrong, a turn of events that he didn't take particularly well. She recalled working alone in the library after Carnegie had emptied out for Christmas 1990, when Sandage burst in in a rage. He had seen her study claiming a higher value. "He was so angry," Freedman later told *Science* magazine, "that you sort of become aware that you're the only two people in the building. I took a step back, and that was when I realized, oh boy, this was not the friendliest of fields."

Today's crisis doesn't appear to have produced quite the same level of acrimony, although tensions, both statistical and personal, are clearly present. Despite their apparent rivalry, Freedman told me she has no doubt that Adam Riess and the SH0ES team are working hard to make the best measurements possible. "I believe he has every intention of doing this in as clear and open way as he possibly can," she said. But that doesn't mean she has total faith in their results.

"Measuring the Hubble constant to 1 percent [the ultimate goal of the SH0ES team] is a tall order," she told me. "Cepheids are these humongous supergiant stars with pulsating atmospheres and luminosities that depend on abundances of elements, plus you have absorption by dust. Trying to get to 1 percent is a similar challenge as it was getting down to 10 percent before Hubble. The question is, are we there? And what do we need to do to get to where we really have nailed it?"

Knowing the pitfalls of Cepheids better than almost anyone, Freedman has sought out alternative ways to measure the Hubble constant, including the opportunities presented by red giant stars that she noted in Santa Barbara. Found at a galaxy's edge, they are situated where astronomers can view them without fear of stuff getting in the way.

Red giants are stars near the end of their lives that have burned through the hydrogen fuel in their cores and expanded to bloated crimson orbs. (This will be the eventual fate of our Sun, several billion years hence, when it will swell to a monstrous size, likely engulfing the inner planets including Earth.) The core of a red giant is made up of helium, the product of the hydrogen fusion reactions that powered the star during its youth. The temperature in the heart of the star is too low to make helium fuse into carbon, so its only source of power comes from a burning shell of hydrogen surrounding the inert helium core. As hydrogen fuses in this shell, helium ash rains down onto the core, gradually increasing its temperature. Eventually, it reaches a blistering 300 million degrees, so hot that helium fusion suddenly ignites, leading to a violent release of energy known as the helium flash.

Now with the core burning helium into carbon, the red giant shrinks back down to a smaller yellow star, which gives off less light overall but has a higher surface temperature.

It is this abrupt change in color and brightness as the red giant phase comes to an end that can act as another standard candle. However, it's not possible to get at distances from a single red giant in the way you can from a Cepheid. Instead, astronomers have to survey a population of red giant stars and look for what is known as the tip of the red giant branch—the stars with the reddest surfaces and the highest light output that are on the verge of shrinking back down to smaller yellow stars. These brightest and reddest stars should all have the same temperature and brightness, regardless of which galaxy you are looking at. So if you can calibrate the total light output of the tip of the red giant branch, you can use it to estimate distances to any galaxy where red giants can be directly observed.

As we saw, when Freedman and her team swapped out the Cepheid-based rung of the distance ladder and replaced it with the red giant method, they got a value of the Hubble constant that landed right in between the direct SH0ES measurement and the standard model prediction. The discrepancy with SH0ES hasn't yet been explained, but it could be a worrying hint that there is something wrong with the Cepheid-based distance ladder.

One thing that keeps Freedman up at night is the huge distances that the SH0ES team is pushing out to with Cepheids now. "As you go out farther and farther . . . crowding effects are going to be a huge problem." What she means is that trying to spy Cepheids in ultradistant galaxies makes it harder and harder to tell them apart from neighboring stars, which could lead to misestimating their brightness. "When you push things to the limit, that's exactly where you run into systematic uncertainties," she warned, "and these kinds of systematics can bite you. Why does this concern me? Well, we've seen this before in the factor-of-two era. It's scary."

Unsurprisingly, Adam Riess disagrees. In Baltimore, he

showed me a graph he'd been working on that compared distances to galaxies estimated using Cepheids and red giants. As far as his calculations go, the two different methods gave more or less identical results. What's more, when the SH0ES team made their own measurement of the Hubble constant using the red giant method, they got a result that agreed pretty well with their preexisting Cepheid analysis.

The fact that Freedman's team got a different result suggests that not all the confounding effects are yet understood. Whether there's a bias in either Riess's analysis or Freedman's is still up for debate.

Freedman is an astronomer to her bones. Until 2020 she spent a lot of her time at the Las Campanas Observatory, high in the Atacama Desert, operating the huge Magellan telescopes. It's a far more romantic experience than getting data from Hubble, which is piped directly to her university office over the internet. "I really enjoy being on the mountain. There's something about just being there that's very different from doing it remotely. Back in the old days we'd sit under the telescope with headphones listening to music, looking up at the sky. You sit in a warm room now, but I still enjoy observing and going down to the mountain."

Since the COVID-19 pandemic, observing has all been done remotely, but Freedman was itching to get back. She also worried about a generation of students who were growing up without ever having touched a telescope. "It's important to have a connection to the instrument, understanding what the conditions were when the observations were made. You lose a lot when people are removed from that."

Freedman was particularly excited for an upcoming generation of instruments, which promise to shed new light on the Hubble crisis. She's told me that she already has observing time booked on the mighty James Webb Space Telescope, the spiritual successor to Hubble that was launched into space on Christmas Day 2021. With a mirror the size of a tennis court made up of eighteen golden hexagonal plates, James Webb is already able to produce images far sharper than Hubble. But its real advantage, at least as far as Cepheids go, comes from the fact that it looks at the universe in infrared, rather than visible light, allowing it to see through dust. Freedman hopes it may allow her and her colleagues to tease out any problems with the distance ladder that have so far eluded them.

While Freedman believes that red giants may ultimately offer a better way of measuring the expansion of the universe than Cepheids, she still has a lot of affection for those young, blinking stars. "I love Cepheids; I spent a huge fraction of my career measuring them. But I also know where the difficulties lie." She resists being characterized as some kind of hardened skeptic: "I don't know, I don't feel particularly skeptical. I'm excited about this problem! We have an opportunity to learn if the standard model is missing something. There's lots of scope for there to be something beyond the standard model: we don't know what dark matter is; we don't know what dark energy is! I don't think I'm pessimistic, I'm just realistic.

"These are challenging measurements," she adds. "I have no idea how this is gonna come out in the end. I think that it would be fascinating if it's new physics. The fact that we're even talking about 1 percent precision, when a few decades ago it was a factor of two, it's fantastic! But are we *really* at the level of 1 percent? Is this *really* a five-sigma result?

"A one-in-a-million chance that this could be wrong?" she wonders aloud. "That sounds optimistic."

The Cosmos Builder

Having worked in Cambridge for most of my career, where the Gothic spires and turrets of the university's medieval colleges dominate the cityscape, I found walking around Princeton a bit uncanny. Much of the campus is built in the Collegiate Gothic style, a twentieth-century imitation of the architecture of Oxford and Cambridge, which lends it the air of an academic Disneyland. One exception is the pleasingly utilitarian Department of Physics, set back from a broad tree-lined avenue that leads up from the town's picturesque canal. I had come to Princeton to meet Jo Dunkley, a world-renowned cosmologist who, like Princeton's architecture, is another Oxbridge import, having studied for her degree in Cambridge and her doctorate at Oxford, before crossing the pond to ultimately become a professor of physics.

Dunkley is an expert on the cosmic microwave background, the primordial light from the big bang that cosmologists use to predict the value of today's Hubble constant. Her work bridges both theory and data analysis (something pretty much unheard of in my own field of particle physics, where people generally fall clearly into either the theoretical or the experimental camp). Dunkley has spent much of her career analyzing data produced by telescopes that study the cosmic microwave background in order to pin down the properties and history of the cosmos.

Her first job after completing her doctorate was as a Princeton postdoc, working on data from the Wilkinson Microwave Anisotropy Probe (WMAP), a forbiddingly named NASA spacecraft that operated between 2001 and 2010. By mapping the cosmic microwave background across the whole sky, WMAP produced a detailed image of the fireball that filled the universe 380,000 years after the big bang. One remarkable thing about the cosmic microwave background is that it has almost exactly the same temperature no matter which direction you look, a remarkably uniform –270.43 degrees Celsius to be precise. Now, that may sound rather chilly for a fireball, but its low temperature is down to the fact that the searing light from the big bang has been stretched by the expansion of space as it traveled through the universe over the past 13.8 billion years, weakening it to a long-wavelength microwave echo.

However, the cosmic microwave background is not *completely* smooth. Measure its temperature with enough precision and you'll find that there are tiny fluctuations down at the level of about one part in a hundred thousand. These minute irregularities are the result of interactions between matter and light in the big bang's fireball and, minuscule though they are, they contain a treasure trove of information about the physics of the very early universe.

As we sat down for a chat in her enviably spacious office, I noticed that (like Riess) she had a beach ball on her shelf with the cosmic microwave background printed on its surface, the primordial fluctuations represented by yellow and blue speckles.

Dunkley's job on WMAP had been to run the last bit of data analysis to extract the parameters of the standard cosmological model: things like the densities of matter and dark matter and the age of the universe. She also tried out alternative cosmological models, where you add in extra particles or exotic types of dark energy, but none of them fitted the data quite as well as the bog-standard cosmological model. "So we came out of WMAP thinking, the standard model looks pretty good," she said. However, while WMAP's data was the most precise available at the time, it wasn't precise enough to rule out alternative cosmological theories; there was still plenty of room for new physics to crop up.

As we've seen, it was the European spacecraft Planck, WMAP's successor, that changed the game. Planck's first image of the cosmic microwave background, released in 2013, was so beautifully detailed that the team was quickly able to rule out a whole host of alternatives to the standard cosmological model. Simultaneously, though, it suggested a Hubble value that put it immediately in tension with the direct measurement produced by Adam Riess and the SH0ES team, kick-starting the whole Hubble tension saga.

Back in 2013, there were some doubts about Planck's finding. They were getting a value of the Hubble constant that was quite a bit lower not just than direct measurements but even than the one calculated from WMAP's data. Since Planck and WMAP were both looking at the same object—the big bang's fireball— they should have been getting the same answer. It was enough to be concerned.

"A bunch of us were thinking, ooh, do we completely trust Planck? Is the cosmic microwave background the problem?"

To tease out any hidden glitches in Planck's data, Dunkley and her colleagues turned to another cosmic microwave background telescope. Perched atop a dormant volcano in the high Atacama Desert in northern Chile is the Atacama Cosmology Telescope, or ACT. The high Atacama is one of the best places on Earth for astronomy thanks to its altitude and clear skies. ACT is especially good at measuring the fine detail in the cosmic microwave background. To see if there was an issue with the Planck data, Dunkley and her colleagues combined data from WMAP and ACT with the aim of remeasuring the Hubble constant completely independently of Planck.

"This time we made sure that we were being good scientists

by blinding our results," she said. By February 2020, Dunkley and her team were ready to open the box. They gathered around the table in her office for the big reveal. Would they get a value close to the Planck prediction—around 67—or one closer to SH0ES, up at around 73. "My colleague Erminia Calabrese, who was connected via Zoom from Cardiff, had run all the models and we unblinded and we were like, 'Oh my God! It's 67! It's the same number!'"

Since then, further measurements of the cosmic microwave background by the South Pole Telescope and alternative methods looking at the clustering of galaxies at grandest scales have firmed up the early universe value of the Hubble constant at around 67.

"So we in the community now agree that the 67 value is robust," Dunkley told me.

A few months earlier, Dunkley had hosted a workshop at Princeton to discuss the Hubble tension. "It was really interesting; we had both Wendy [Freedman] and Adam [Riess] presenting," she explained. "It was pretty clear that there isn't yet agreement in the local measurements on what the systematic error budget is, and there were, umm, some *useful* discussions between the different groups about that." Asked what she meant by "useful," she replied with a wry smile, "Well, let's just say the Hubble constant has long had this history of tensions between different groups, and it hasn't gone away."

All of which brought us to the real point of my visit. As a somewhat independent observer of the controversy over the direct measurements, one with a background in theory and data analysis alike, what did Dunkley make of the current situation?

"The Hubble tension is super interesting, but I'm not yet convinced that it's enough to break the standard model of cosmology." To be convinced, she said, "we need the direct measurement gang to agree with each other."

And what if the Hubble tension *is* real? What could it be telling us about the universe? The other prong of Dunkley's work is testing alternative cosmological models, to see if they could explain the tension. It's fair to say that the theoretical community has so far found it hard to come up with alternatives that don't introduce tensions of their own. In fact, it's the absence of a compelling theoretical explanation for the Hubble tension that causes some cosmologists to suspect it really is just down to mistakes in the measurements.

Dunkley doesn't agree with that way of thinking. "The fact is you can't argue, in the end, with data. However much the theorists say, 'Oooh, it's quite hard to do,' if you have to do it, you have to do it. So, we shouldn't really pay much attention to it being hard. Whatever nature is really telling us, we have to find a model that works."

It could be, for instance, that an exotic form of dark energy drove the universe to expand faster during its early history. This so-called early dark energy would have permeated the cosmos during the first 380,000 years of cosmic time, at a time when the universe was filled by the burning primordial fireball from the big bang. Then, as the fireball cooled and faded to produce the cosmic microwave background, this early dark energy also vanished, leaving no trace in the later universe. Or it could be that neutrinos, so hard to pin down today, were somewhat less ghostly and could bump into each other during the hot, dense early universe, in a way that affected the rate of cosmic expansion.

"None of them get up to 73, but you can squeeze toward it," Dunkley said.

What really excites Dunkley is the imminent prospect of testing these early dark energy models directly. Early dark energy would leave a noticeable signature on the cosmic microwave background, and it should soon be possible to put it to the test with the latest data collected by the Atacama Cosmology Telescope. A whole program is planned to gradually improve our understanding of the cosmic microwave background, including the construction of the Simons Observatory, currently being built on the same Chilean volcano as ACT.

Just three days before we spoke, President Biden had unveiled the first image produced by the James Webb Space Telescope. It revealed a universe teeming with thousands of galaxies, all squeezed into a tiny patch of sky about the size of a grain of sand held at arm's length. Some of these galaxies formed a mere billion years after the big bang and are among the faintest and most distant objects ever observed.

It is incredible instruments like Webb that will ultimately allow Dunkley, Riess, Freedman, and their thousands of colleagues to continue to decode the universe's secrets. Soon enough they may tell us if the Hubble tension is real.

Dunkley began her career in 2001, in the aftermath of the discovery of the accelerating universe and dark energy. After a century of breakthroughs that rewrote the cosmic story, though, there hasn't been a major change in how we think about the cosmos since. But despite the gloominess that has taken hold of some elements of the physics community, she doesn't think that the age of big breakthroughs and revolutions is over.

"I don't know when the next breakthrough is going to come, but the Hubble tension might be the clue we're after," she said with a smile. "I'm all agog!"

Chapter 9

Space Oddities

Anomalies rise and anomalies fall,
and new anomalies rise again.

As scientists grapple with long-standing puzzles like the Hubble tension and the peculiar magnetism of the muon, new and strange phenomena are constantly being uncovered. This embarrassment of scientific riches is what makes the pursuit of anomalies so endlessly fascinating, helping us refine both how we do science and our understanding of nature. So, before we draw our story to a close, here are a couple more emerging enigmas that may prove significant in the coming years.

What's Eating the Universe?

Alexandra Amon is a young cosmologist at Cambridge's Kavli Institute for Cosmology, a group of pavilions set among leafy parkland on the edge of the city. She is at the center of an anomaly story that, like the Hubble tension, is threatening the standard cosmological model.

Amon grew up on the tiny Caribbean islands of Trinidad and Tobago before moving to the U.K. to study at the University of Edinburgh. There she worked on the Canada-France-Hawaii

Telescope Lensing Survey (CFHTLenS), an international project that used a huge telescope perched on the summit of Mauna Kea, a dormant volcano in Hawaii, to map dark matter. You might very well wonder how a telescope can be used to map dark matter, given that, famously, it's completely invisible. Dark matter does, however, leave its mark on the visible universe, in particular through an effect known as gravitational lensing.

According to Einstein's general theory of relativity, gravity is not a force but a manifestation of the curvature of space and time. Massive objects like stars and galaxies bend the fabric of space-time around them, like a bowling ball plonked on the surface of a trampoline. As a result, when light passes close to massive objects, its path, too, must be bent as it travels through the curved space. This means that anything with enough mass, say a big cloud of dark matter, can act like a lens, distorting how visible galaxies and stars appear in the night sky.

Surveying around ten million galaxies, the CFHTLenS team found that many looked as though they had been slightly squashed in one direction or another as their light was lensed by dark matter on its way to Earth. By mapping where these distortions were greatest, the team was able to reveal vast web-like structures of dark matter threading their way through the cosmos at the grandest scales imaginable. This enormous cosmic web is the invisible skeleton on which our visible universe hangs. More than that, its gravity was responsible for drawing matter together after the big bang to form the first galaxies and stars.

However, the final results of the CFHTLenS project suggested something unexpected: these vast structures were less clumpy, for want of a better word, than the standard cosmological model predicted. As with the Hubble tension, measurements of the cosmic microwave background allow cosmologists to determine the properties of the early universe, in particular how evenly or unevenly matter was distributed in the fireball following the big bang. Then, using the standard cosmological model, they can run the clock forward to predict how these primordial seeds should have grown as their gravity pulled in more and more matter to form the giant structures we see in the universe today.

The problem is, these predictions suggest that the universe ought to be denser and more clumpy than it appears. Or to put it another way, like a watery soup, the universe is simply too thin.

Initially, cosmologists were skeptical of the team's result, but during Amon's PhD more and more observations began to confirm their findings. As Amon told me when we met, "I grew up hearing that CFHTLenS was wrong. Then the story kind of changed and people were like, 'Huh, lensing surveys are all low, not just CFHTLenS.' So then it changed from just CFHTLenS being wrong to *all* lensing being wrong."

Amon now leads the lensing team working on the Dark Energy Survey, a major international project that spent six years mapping more than 300 million galaxies in an attempt to better understand how dark energy shapes the universe. When they unblinded their results, at the end of 2020, they found that they too were in tension with the predictions of the standard cosmological model. As with many of the results we've seen, their tension was only at around the two- to three-sigma level, still far below a formal discovery. But taken together with other results, they seemed to be strongly pointing to *something* strange going on in the cosmos.

In 2021, Amon moved to Cambridge to work with George

Efstathiou, one of the world's most eminent cosmologists and an architect of the standard cosmological model. Efstathiou is an arch skeptic when it comes to the Hubble tension and has engaged in a long running debate with Adam Riess, arguing that it's far more likely to be the product of missed experimental effects than the harbinger of the fall of the cosmological model. When Amon arrived in Cambridge, he had a similar attitude toward the new tension over the clumpiness of the universe. That was at least part of why she wanted to work with him.

However, a year and a half on, she told me that he's been won over. So much so that they recently put out a paper together in an attempt to explain what could be making the universe more spread out than we'd expect. There were two broad explanations.

The first is to do with the most powerful objects in the universe: supermassive black holes. These behemoths are found at the centers of galaxies and can weigh tens of billions times more than our Sun. Their enormous gravitational pull draws gas and dust from the surrounding area into glowing disks of superheated matter that shine with the light of countless stars. As the black hole devours material from the disk, it can produce indescribably powerful jets of energy that blast outward across thousands of light-years and heat any surrounding gas to high temperatures, which has the knock-on effect of preventing stars from forming in any region unlucky enough to end up in their firing line. Supermassive black holes therefore block structure formation in the universe, making it less clumpy overall. Astrophysicists have done their best to estimate how big such an effect may be, but there is a chance that it's much larger than thought.

That would be the "boring" explanation for the anomaly,

but frankly it would still be pretty exciting to learn more about these incredible objects. The *interesting* explanation, on the other hand, relates to the nature of dark matter.

The standard cosmological model assumes that dark matter is made up of a single type of massive particle. However, this rather simple picture need not be true. It could well be that there are multiple types of particles in the dark universe, just as our visible world is made up of several fundamental ingredients. This is what Amon and Efstathiou's work seems to be hinting at. She explained, "I think it's on the dark matter side; I think there's going to be some paradigm shift. When I first heard about multicomponent dark matter, I just thought, 'That's even more weird; why would that be a thing?' But I've changed my mind now. Even George [Efstathiou] thinks it's more likely to be something to do with dark matter."

New data will soon allow Amon and her colleagues to settle the debate. She's particularly excited for two telescopes that are in the works. First is the Vera C. Rubin Observatory, a large new telescope, also on a Chilean mountaintop, named after the astrophysicist who found the first solid evidence for the existence of dark matter. Rubin will be a game changer thanks to its ability to map a billion galaxies in a single year. However, it will really come into its own when combined with data from the European Space Agency's Euclid spacecraft, an infrared telescope that will be able to study galaxies outside the obscuring haze of Earth's atmosphere.

Together, these two observatories will change the game. As Amon told me with evident exhilaration, "The bombardment of data on the cosmology side means we're not just going to be sitting on two to three sigma; we're going to be sitting on huge tensions!"

The Biggest Bang Since the Big One

On October 9, 2022, at 14:10 GMT, an almighty burst of radiation ripped through the solar system and swept over the Earth. Orbiting spacecraft designed to detect gamma rays, temporarily blinded by the intensity of the pulse, quickly relayed information on the event to astronomical observatories around the world. As telescopes trained their attention on its source, it soon became clear that we had just witnessed the most powerful cosmic explosion in recorded history, perhaps even since the dawn of civilization.

The blast was what astronomers call a gamma-ray burst, the most potent class of explosions in the known universe, more brilliant than anything that has occurred since the big bang itself. Formally, this cataclysmic event entered the astronomical history books as GRB221009A,* but it quickly acquired a catchier nickname: "the BOAT," meaning "the brightest of all time." Astronomers have calculated that even if you converted the entire mass of the Sun into pure energy, that still wouldn't come close to the unimaginable power of the BOAT.

The nature of gamma-ray bursts is still uncertain, but astronomers believe that they are caused when a giant star goes supernova, imploding to form a black hole. As the black hole is born, it consumes vast quantities of the infalling stellar matter, creating powerful jets of radiation that flare out in opposite directions, like the beams of a terrible lighthouse. When one of these death rays is pointed at Earth, we witness an intense flash of gamma rays. Fortunately for us, the BOAT occurred a long

---

* "GRB" is for "gamma-ray burst," the numbers are the date in YY-MM-DD format, and the *A* at the end means it was the first gamma-ray burst that day.

time ago in a galaxy far, far away—around 1.9 billion years ago in fact—but that still makes it one of the closest such explosions ever detected.

When astronomers studied its radio afterglow, they found that its behavior didn't match what you'd expect from a black hole born of a supernova. As radiation from the imploding core breaks through the outer body of a dying star, astronomers expect to see sudden jumps in the brightness of the burst. Instead, the light from the BOAT seemed to rise and fall smoothly. This has led some astronomers to wonder if we really understand gamma-ray bursts at all. And if it wasn't a supernova that caused the brightest explosion of all time, then what was it?

All that is remarkable enough. But an observatory high in the mountains of Sichuan, China, found something even more peculiar about the radiation it had emitted. The gamma rays from the explosion had energies far above what our current theories tell us ought to be possible, with a single photon carrying more than twice the energy of a particle accelerated by the Large Hadron Collider, the world's highest-energy accelerator. What's puzzling is that such photons should never make it far enough across the cosmos to reach us here on Earth. As they travel through intergalactic space, these super-energetic gamma rays should collide with lower-energy photons from stars and galaxies and convert into electrons and antielectrons in the process. So how did they get here?

A tantalizing possibility is that these high-energy gamma rays actually originated from a speculative form of dark matter particle that physicists call an axion. Blasted out from the black hole at tremendous speeds, these axions would have been able to zip through intergalactic space unhindered before converting into photons when they passed through a strong magnetic field somewhere closer to Earth, explaining how such a high-energy particle made it such a vast distance across space.

If that's right, then it would be a huge breakthrough, as the theoretical physicist Ian Shoemaker told me. Having worked on theories to explain the impossible particles emerging from beneath Antarctica detected by the ANITA balloon experiment, Shoemaker said this new cosmic anomaly captured his attention both for its scientific potential and for what it told us about how science progresses. "One bit of evidence of how crazy the field is about these things: it only came out in October, and [a month later] there's not even an experimental paper about this yet, but there are already a load of theoretical explanations."

Shoemaker for one relishes this anomaly-led era. "It's a fun time in a lot of ways because there's so much activity in a lot of different directions, but also an unsettling time because there's no hegemonic theory right now that's telling us where we should really be searching." A common criticism of theorists engaged in such work is that they are "ambulance chasers," rushing to the site of disasters in search of new business. "I understand that critique," Shoemaker said, "but even when anomalies fail and there's no fundamental breakthrough, they really motivate people to break out of the narrow theoretical frameworks that they're operating in.

"Ultimately, I think they're great for the community, for both experimentalists and theorists," he added. "They motivate excitement and interest."

Chapter 10

The Unsearchable
Riches of Creation

n hour's drive west from Princeton is the picturesque township of Holmdel, New Jersey. As I wound
my way along its wooded roads, I was occasionally
rewarded with views across wide open lawns to generous clapboard houses, Stars and Stripes fluttering from fence posts in
the summer breeze. It struck me as the kind of quintessential
small American town that Bill Bryson went searching for in
his first travel book, *The Lost Continent:* tranquil, affluent, at
ease with itself. All that was missing were some white picket
fences. However, despite Holmdel's obvious charms, I hadn't
come there to look for America. I was on a scientific pilgrimage
to see the place where one anomaly changed everything.

On the edge of town, I found what I was looking for, a right
turn marked "Bell Labs." A sign warned sternly, "No Trespassing, All Visitors Must Report to Main Reception." But it was a
Saturday, and the main reception was locked and dark. Against
all my rule-abiding instincts, I decided to press on, tentatively
crawling up toward the parking lot, being sure to at least obey
the ten-mile-per-hour speed limit. Parked under a cool canopy
of trees, I watched the rearview mirror for a few moments, half

expecting to see a security guard striding up the hill to give me a telling off. But when no one appeared, I got out and walked up to a wide steel gate barring the road. Another sign warned against trespassing, but I stepped undeterred around the padlocked gate and walked through the trees until the wood opened onto a green glade.

At the far end, I saw it, an image straight out of the books and lectures of my youth: the Holmdel Horn Antenna.

Up close, the antenna is an imposing if rather unprepossessing thing, an aluminum horn the size of a house mounted upon a large steel turntable that once allowed its users to point it anywhere in the sky. When it was built in the late 1950s, its steel frame was painted gleaming white, but today the paint is peeling and flecked with orange rust. A stone plinth reports that it was designated a national monument in 1989. I hope it's being taken good care of, because for my money there can hardly be a more significant monument anywhere on Earth.

In 1964, two American radio astronomers were working with the Holmdel Horn when they stumbled upon an anomaly that would change how we see the world forever. Arno Penzias and Robert Wilson were planning to use the antenna to make radio observations of the Milky Way, but as they were preparing the equipment, they found that the horn was picking up a faint microwave buzz. Realizing that the noise would wreck their astronomical observations, they set about trying to track down its source and eliminate it. At first they wondered whether they might be getting interference from stray radio broadcasts from New York City, which was just sixteen miles away across Sandy Hook Bay. However, when they pointed the antenna directly at the city, or indeed in any other direction, the microwave noise remained irritatingly constant. Another hunch, that it might be emanating from the Milky Way, turned out to be wrong too. Eventually they were forced to the conclusion that the noise must be coming from the horn itself. Climbing inside, they discovered that two pigeons had taken up roost where the horn narrowed and entered the warm observation cabin. As Robert Wilson later put it, "They had covered the inside with a white material familiar to all city dwellers," but after they evicted the squatters and cleaned up their mess, the buzz remained stubbornly persistent.

A year went by without a solution. In the spring of 1965, after performing what observations they could, Penzias and Wilson gave the whole horn a thorough clean, putting aluminum tape over the riveted joints and even dismantling and reassembling the narrower section, all to no avail.

A while later, Penzias was at MIT when he got talking to the astrophysicist Bernard Burke, telling him about the trouble they were having at Holmdel. The story reminded Burke of some work that the astronomers James Peebles and Robert Dicke had been doing on a model of the universe that went through an extremely hot, dense phase. They had calculated that if the universe had once been much smaller, denser, and hotter, there should be a microwave glow coming from the whole sky. Penzias and Wilson reached out to Dicke, and he and his colleagues came to Holmdel to meet the pair and inspect their equipment. Seeing the care that they had taken with their measurements, Dicke was convinced that their results were solid.

It soon became clear: the noise that had plagued the pair of astronomers was nothing less than the faded light of the primordial fireball that once filled the universe. Penzias and Wilson had stumbled upon the afterglow of the big bang.

The effect of their discovery was profound. It finally settled the debate over the universe's origins, one that had raged since Edwin Hubble discovered that space was expanding, more than three decades earlier. The universe was not eternal; it had a beginning. The signal they had picked up came from the entire sky because the big bang happened everywhere, all at once. The discovery of the cosmic microwave background, as it is now known, gave birth to observational cosmology. Arno Penzias and Robert Wilson's microwave buzz is now the prime tool of cosmologists seeking to understand the universe and its origins.

And all that from one apparently inconsequential anomaly.

Will any of the anomalies we've encountered have as profound an impact as Penzias and Wilson's? That is, of course, a very high bar. You can only discover the big bang once. But there's certainly a chance. There are intriguing hints that some of these disparate results could be related to one another, revealing different glimpses of a single larger reality. Several attempts have been made to find common explanations for multiple anomalies at once. For instance, theorists have managed to combine the hints we've been getting from beauty quarks at LHCb with the anomaly in the muon's magnetism detected at Fermilab to trace the outline of a new fundamental force. Similarly, when I spoke to Adam Riess about the Hubble tension, he told me that if a new type of neutrino turned up in experiments, that would be game over, proving the Hubble tension was the real deal once and for all. Neutrino experiments seem to be revealing hints of exactly that. Could they be connected?

Maybe, maybe not. Recall Adam Riess's old professor who would proclaim when his eager students brought him news of the latest anomaly, "It's wrong." Anomalies come and go, and most of the time, as we've seen, they *are* wrong. I know it too

well, having watched convincing evidence for a new fundamental force evaporate before my own eyes.

But the wealth of different anomalies, in particle physics
and cosmology, gives me hope that we will soon find the crack
that will allow us to prize open a window to a new view of
nature. I find the anomalies in cosmology particularly compelling. While the standard model of particle physics is fantastically mature and well tested, the standard cosmological model
rests on shakier, more novel foundations. The fact that we have
no idea what its two main ingredients—dark energy and dark
matter—are leaves a huge amount of room for new discoveries to be made. The Hubble tension and the problem over the
clumpiness of the universe seem to be pointing strongly to the
fact that the standard cosmological story is ripe for rewriting.

Even when such anomalies go wrong, we learn something
from them. They are the grindstone on which our scientific
tools are sharpened. Pursuing them refines our experimental
techniques and deepens our theoretical understanding. An
anomaly that doesn't lead to a revolutionary new discovery can
as often prepare the way for the next big breakthrough.

In 1871, the Scottish physicist James Clerk Maxwell gave a
lecture to set out his vision for what would become the Cavendish Laboratory in Cambridge, where I now work. After enormous progress in understanding nature during the nineteenth
century, a view had gotten about that the greatest discoveries had already been made. All that remained for science was
the workmanlike task of measuring physical constants to ever
greater precision. You often hear similar views being expressed
today—that physics has reached a dead end, that there is little
room left for progress. Maxwell, widely regarded as ranking
alongside Newton and Einstein for his foundational work in

theoretical physics, took issue with this view in a beautiful peroration:

> We have no right to think thus of the unsearchable riches of creation, or of the untried fertility of those fresh minds into which these riches will be poured. It may possibly be true that in some of those fields of discovery . . . the great explorers of former times have appropriated most of what is valuable. . . . But the history of science shows that even during that phase of her progress in which she devotes herself to improving the accuracy of the numerical measurement . . . , she is preparing the materials for the subjugation of new regions, which would have remained unknown if she had been contented with the rough methods of her early pioneers.

Though couched in rather Victorian terms ("subjugation of new regions" hasn't aged well), Maxwell's point was well made. He wouldn't live to see his faith vindicated, but by the end of the century the very lab that he was founding became the center of a revolution that transformed our understanding of nature beyond recognition, as atoms were torn apart and new forms of energy released. Physics was far from done in 1871.

Nor is it done now. Perhaps the greatest change in fundamental science over the past decade has been the end of the theory-led era. Since the 1970s, particle physics has been dominated by the incredibly successful standard model. For decades it seemed as though the only job of experimental physics was to confirm the predictions of this magnificent theoretical edifice, a process that reached its acme with the discovery of the Higgs boson in 2012, the theory's last unconfirmed prediction. However, since then, none of the ambitious theoretical ideas—be they supersymmetry or grand unified theories—that promised to resolve the many remaining mysteries have stood up to experimental scrutiny. Finding ourselves in unknown territory, without a theoretical map to guide us, has bewildered and disheartened many.

But perhaps it was that earlier, theory-led period that was the real anomaly. Historically, the vast majority of the great breakthroughs in our understanding of atoms or the evolution of the cosmos were driven by experimental discoveries. The grand theorizing usually came only after the fact. Today we are once again off the edge of the map, where experimental clues must be our lodestars.

We do not yet know how this story will end. But that is what it is to do science: to set out on a journey with no guarantee of success, without knowing where the road will take you, or even if you have set out in the right direction. Science does not progress in a straight line, running from ignorance to understanding. It is a messy business, full of false starts, wrong turns, and dead ends. Nature does not yield its secrets easily; they must be fought for. But in the end, despite the setbacks and the false dawns, this winding road does inexorably lead to deeper understanding.

I hope that the stories in this book have given you a flavor of what it is like to work at the cutting edge of science. Whether these anomalies disintegrate or, like Penzias and Wilson's insistent buzzing, form the bedrock of new physics, they have already taught us so much: how the hearts of supergiant stars beat, the way the strong force pulls on the smallest building blocks of nature, and even what lies beneath the Antarctic ice sheet.

Of one thing I'm sure: sooner or later, someone, perhaps trawling through their latest experimental data in a dingy office late one evening, or sitting under the dome of a mountaintop observatory staring up at the shimmering night sky, will notice something strange. They may not realize it, but they will have discovered a key to a new world. Perhaps in that moment, as they struggle to make sense of what they've seen, they'll find themselves uttering science's most auspicious phrase: "Hmmm, that's funny . . ."

Acknowledgments

Writing this book wouldn't have been possible without the help and expertise of numerous people who gave their time so generously to share their work and their extraordinary stories and to show me around their laboratories. I'd particularly like to thank Chris Polly for giving up an entire day to offer me a tour of Fermilab and the marvelous Muon $g - 2$ experiment, and Mitesh Patel and Vava Gligorov for their thoughtful and constructive feedback. I'm also very grateful to Nicola Rossi, Linda Cremonesi, Ryan Nichol, Ian Shoemaker, Mark Lancaster, Christine Davies, Bonnie Fleming, Joachim Kopp, Matt Toups, Nico Serra, Paula Álvarez Cartelle, Ben Allanach, Gino Isidori, Gudrun Hiller, Adam Riess, Blake Sherwin, Eleonora Di Valentino, Jo Dunkley, Wendy Freedman, and Alexandra Amon.

Beyond the names that appear in this book are countless other scientists, engineers, technicians, and support staff who make today's great voyages in particle physics and cosmology possible. Without them I'd have nothing to write about. Thanks also to the always-helpful librarians at the Science Museum Library and Archives and the Rayleigh Library, to Mel Northover for turning my amateurish scrawlings into far more attractive diagrams, and to Ingrid Sterner for her careful copyedit.

As ever, I'd like to thank Val Gibson for her support and encouragement. I owe a special debt of thanks to my brilliant agents Simon Trewin in London and Dorian Karchmar at WME in New York. Thank you for making this adventure in writing possible. I'm deeply grateful to my editor Yaniv Soha at Doubleday for his insight, guidance, and encouragement throughout, and to Ravi Mirchandani at Picador for backing me from the start.

Finally, I'd like to thank my friends and family for their love and support, especially when the going was tougher. Robert, Vicky, Alexandra, and Alastair—I'm so lucky to have you in my life.

Notes

Chapter 1: The Cosmic Story

20 0.00115965218161: T. Aoyama, T. Kinoshita, and M. Nio, "Theory of the Anomalous Magnetic Moment of the Electron," *Atoms* 7, no. 1 (2019).

20 0.00115965218059: X. Fan et al., "Measurement of the Electron Magnetic Moment," *Physical Review Letters* 130 (Feb. 13, 2023): 071801.

Chapter 2: The Fall of Planet Vulcan

36 "I have now derived": *The Collected Papers of Albert Einstein*, vol. 8, document 144a (Princeton: Princeton University Press, 1987).

Chapter 3: You Are the Easiest Person to Fool

44 "first tremors of the Big Bang": Brian Keating, *Losing the Nobel Prize* (New York: W. W. Norton, 2018), 200.

44 "the greatest discovery of the century": Ibid., 201.

Chapter 5: A Magnetic Mystery

90 "Muons are weird": Will Kinney (@WKCosmo), Twitter, Aug. 10, 2022, 6:00 p.m., www.twitter.com.

91 "Who ordered that?": Martin Bartusiak, "Who Ordered the Muon?," *The New York Times*, Sept. 27, 1987.

94   "achingly beautiful": Frank Wilczek, Paul Dirac Centenary Symposium, Florida State University, Dec. 2002.

96   "half genius; half buffoon": Frank Close, *The Infinity Puzzle* (Oxford: Oxford University Press, 2011), 32.

97   "little squirt": Richard Feynman, interview by Charles Weiner, June 27, 1966, Niels Bohr Library & Archives, American Institute of Physics, www.aip.org.

98   "Is it unitary?": Jørgen Veisdal, "When Feynman Met Dirac," *Cantor's Paradise*, Feb. 26, 2020, www.cantorsparadise.com.

104    $g = 2.00231930436321$: Aoyama, Kinoshita, and Nio, "Theory of the Anomalous Magnetic Moment of the Electron."

104    $g = 2.00231930436118$: Fan et al., "Measurement of the Electron Magnetic Moment."

118   "My wife won't pick me": Adrian Cho, "The Cloak-and-Dagger Tale Behind This Year's Most Anticipated Result in Particle Physics," *Science*, Jan. 27, 2021, www.science.org.

124   wangle time on supercomputers: Natalie Wolchover, " 'Last Hope' Experiment Finds Evidence for Unknown Particles," *Quanta Magazine*, April 7, 2021, www.quantamagazine.org.

125   6 percent of its total data set: Fermilab, "First Results from Fermilab's Muon $g - 2$ Experiment Strengthen Evidence of New Physics," April 7, 2021, news.fnal.gov.

Chapter 6: Ghosts in the Machine

148   "We had these yoga mats": Thomas Lewton, "Is the Great Neutrino Puzzle Pointing to Multiple Missing Particles?," *Quanta Magazine*, Oct. 28, 2021, www.quantamagazine.org.

Chapter 7: Beauty and Truth

175   500 million stampeding horses: Marco Calviani, "Intercepting the Beams," *CERN Courier*, May 3, 2021, www.cerncourier.com.

176   a winter thunderstorm knocked out power: Rende Steerenberg, "LHC Report: Another Run Is Over and LS2 Has Just Begun . . . ," Dec. 11, 2018, www.home.cern/news.

Chapter 8: War in Heaven

224 558 kilometers per second per megaparsec: "Hubble Completes Eight-Year Effort to Measure Expanding Universe," May 25, 1999, www .hubblesite.org.

226 once every five hundred years in the Milky Way: "Type Ia Supernovae," Nancy Grace Roman Space Telescope, www.roman.gsfc.nasa.gov.

240 "He was so angry": Joshua Sokol, "A Recharged Debate over the Speed of the Expansion of the Universe Could Lead to New Physics," *Science*, March 8, 2017, www.science.org.

250 a grain of sand held at arm's length: Rob Garner, "NASA's Webb Delivers Deepest Infrared Image of Universe Yet," NASA, July 12, 2022, www .nasa.gov.

Chapter 9: Space Oddities

256 the entire mass of the Sun into pure energy: Leah Crane, "The Most Powerful Space Explosion Ever Seen Keeps Baffling Astronomers," *New Scientist*, March 28, 2023, www.newscientist.com.

257 1.9 billion years ago: Francis Reddy, "NASA Missions Study What May Be a 1-in-10,000-Year Gamma-Ray Burst," NASA, March 28, 2023, www.nasa.gov.

Chapter 10: The Unsearchable Riches of Creation

261 "They had covered the inside": Robert W. Wilson, "The Cosmic Microwave Background Radiation," Nobel Lecture, Dec. 8, 1978.

264 "We have no right": J. G. Crowther, *The Cavendish Laboratory, 1874–1974* (New York: Science History Publications, 1974), 38.

FUGITIVE'S
TRAIL

Also by Robert J. Conley
in Large Print:

Crazy Snake
The Way South
Go-Ahead Rider

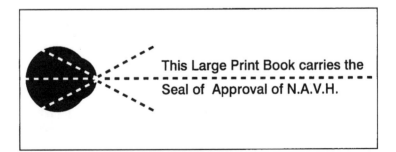

This Large Print Book carries the
Seal of Approval of N.A.V.H.